Understanding Effects Across Space

Extended Magnetism Edition

Understanding Effects Across Space

Electromagnetism, Gravity and Inertia

Extended Magnetism Edition

George Henry Edwards

Copyright © 2017 by George Henry Edwards
Permission is granted for anyone to quote or copy and paste any or all of the copyrighted material herein if the book title: *Understanding Effects Across Space* or the URL: www.giac2002.org/understanding.html is cited as the source.

LIMIT OF LIABILITY/DISCLAIMER OF WARRANTY

The author makes no representations or warranties with respect to the accuracy or completeness of the contents of this work. The fact that a work, person, organization or Website is referred to in this work as a citation or potential source of further information does not mean that the author endorses the information the work, person, organization or Website may provide. Readers should be aware that Internet Websites listed in this work may have changed or disappeared between when this work was written and when it is read.

Library of Congress Control number: 2017915001
ISBN-13: 978-1975953409
ISBN-10: 1975953401
CreateSpace Independent Publishing Platform
North Charleston, SC

Dedicated to Marian Weiss Edwards

Acknowledgments

Thanks to Dr. Amitabha Ghosh for his kind advice and help with especial regard to his comments and corrections to my attempted explanations of his unified theory of gravity and inertia and to Susan Edwards Gum, in addition to her cover art, graphics, and book design, for her valued meticulous copy-editing and advice for both the printed and Web book versions.

My gratitude to my loving wife Marian for reading this book as it was written chapter by chapter and offering comments and advice especially towards helping me make my wording clear.

Table of Contents

Understanding Effects Across Space
Extended Magnetism Edition

Preface — xv

    Introduction — xv
    Using This Book — xvi
    Optional Web Links — xvii

Chapter 1 Understanding the Foundations — 1

    General take on Science, Logic and Mathematics — 1
    Systematization of Logic and Mathematics Particularly in Science — 2
    General Take on the Physical Sciences as Applied to Engineering — 2
    Classical Mechanics — 4
    Maxwell's Equations — 4
    Einstein's Theory of Special Relativity (ESR) — 4
    Einstein's Theory of General Relativity (EGR) — 6
    The Standard Theory of Quantum Mechanics — 8
    Superstrings and Physical Models — 9
    Electric Charge, Magnetism and Electromagnetic Fields — 9
    World Wide Web Lectures — 10
    Conclusions — 11

Chapter 2 Mathematics — 13

    Introduction — 13
    Counting, Numbers and Arithmetic Processes — 15
    Symbols and Language — 18
    Euclid's Plane Geometry — 19
    Trigonometry — 20
    Algebra — 20
    Coordinate Systems, Mappings and Analytical Geometry — 23
    Calculus Terms — 25
    Calculus — 27
    Real Mathematical Analysis — 29
    Vectors — 30
    Imaginary Numbers, Complex Numbers, the Complex Plane and Hyper Complex Numbers — 31

Quaternions	32
Tensor Analysis	33
Probability	34
Differential Equations and Transform Methods	34
Other Geometries	34
Boolean Algebra, Digital Logic, Digital Arithmetic & Digital Computers	35

Chapter 3   Electricity and Magnestism with Hydraulic Analogies — 39

Introduction	39
Physical Fields	39
Hydraulic Analogies	41
Hydraulic Analogy to Electron Flow and Resistance	42
Hydraulic Analogy to Electrical Capacitance	46
Capacitance	46
Magnetism	44
History	48
Magnetic Force	48
Electromagnetism	49
All Magnetic Force is Caused by the Motion of Electric Charge	51
Basic Explanation for Magnetism with Einstein's Special Relativity	53
Is it Best to Name a Magnetic Field Separate from a Moving Coulomb Field?	55
Electron Spins and Orbits may be Positioned in an Object to Make Miniature Magnetic Domains and Produce a Magnetic Force	56
Magnetism Explained in a Nutshell	57
Magnetic Effect on a Magnetizable Object of a Current Carrying Helix without a Ferromagnet Core	57
Nomenclature	58
Hydraulic Analogy to Magnetic Force	59
Qunatification and Practical Applications	59
Inductance	60
Inertia, Induction and Inductance	60
Self Induction or Inductance	61
Inductance as a Circuit Element	62
Hydraulic Analogies for Electricity and Inductance	63

Hydraulic Analogy for Magnetism	64
Transformers	65
Hydraulic Analogy to Electromagnetic Fields	66
Hydraulic Analogy to Electromagnetic Wave Propagation	66
Alternating Current Effects	67
Tuned Circuits	68
Electric Motors and Generators	68
Hydraulic Analogies to Individual Electrons	69
Active Electronic Devices	70
Electrical Circuits	71
Other Electricity Uses	71
Overall Perspective on Electricity	72
Maxwell's Equations: A Unified Theory of Electricity and Magnetism	73
Maxwell's Equations	73
Displacement current	75
Maxwell's Equations Collectively	76
Directions of Field Lines and Return to Hydraulic Analogy	78
Antennas	80
Vectors and Mathematical Tools in Electricity other than basic Algebra and Calculus	81
Basic Alternating Current Circuit Equations	81
References with Comments	81

Chapter 4   **Feedback**   83

Introduction	83
Servo-mechanisms	84
Feedback Resistors	84
Digital Feedback	85
Fire Control Radar Range and Angle Tracking Feedback	85

Chapter 5   **Electronic Design Hardware and Software**   87

Very Basic Electronic Logic	87
Computers	89
Reference	91

Chapter 6	**Where To From Here and Now in Space and Time, The Physics Odyssey**	93
Chapter 7	**Unified Theory of Gravity and Inertia**	101
	References with comments	105
Chapter 8	**Putting It All Together on Gravity and Inertia**	107
	Putting it all together	107
	Application of Einstein's Theory of General Relativity	108
	Conclusion	110
	Addendum—What is Fundamental	111
Appendix	**Mathematics and Individual Equations**	113
	Description of the Mathematical/Physical Terms used in Maxwell's Equations	113
	Verbal Descripton of Each Equation	115
References		117

Table of Contents
Quantum Electrodynamics Sequel

Introduction		121
Chapter 1	**Knowledge, Truth, Trust, Proof, Belief and this Book**	123
Chapter 2	**Distillation of Quantum Electrodynamics [QED]**	127
Chapter 3	**The Beginnings**	129
Chapter 4	**Probabilities and Waves**	131
	Probability Principles Adequate for Quantum Physics Explanations	131
	Complex Numbers	132
	Wave Reprentations	134
	Water and Electromagnetic Waves	134
Chapter 5	**QED Theories**	137
Chapter 6	**Feynman-Path Integral Approach**	139
Chapter 7	**Synopsis: The Book QED, *The Strange Theory of Light and Matter***	147
Chapter 8	**Synopsis: The Book *The Quantum Universe***	149
References		153

Preface

I was, I guess, about 12 years old when I somehow procured [Dad gave me?] a horseshoe magnet that I found would pick up a nail across apparently empty space. Sometime later, I wrapped insulated wire around an iron railroad spike and connected the ends of the wire to a dry cell battery to make an electromagnet. This all led to my interest in understanding effects across space including magnetism, electrical effects and physical science in general. This book summarizes the results. I hope that you find these results of my quest as interesting as I have.

There is a joy in understanding. This book is directed towards understanding rather than application which leads to its own joy—accomplishment. It explores those things that are not directly perceivable to our senses—actions across apparently empty space and electricity.

The foremost aim of this *Enhanced Magnetism Edition* is to provide better understanding of the magnetic force.

It has been known for a couple of centuries that magnetic effects are always due to moving electric charges. So I consider it a potential hindrance to understanding to continue to use the historical term magnetic field when moving electric charges are really what are involved with magnetic effects. I have chosen to often refer to the resultant field as an MCF *field* [MCF standing for *moving Coulomb force*] rather than the historic term, magnetic field.

Such usage also avoids the prevailing association of the Coulomb force strictly with static charges. Charles-Augustin de Coulomb published in 1784 what has since been called *Coulomb's law* that the force between charges is proportional to the product of the charges divided by the square of the distance between them.

I hold any specific mathematics to a bare minimum. I have a reputation for honesty and have nothing to gain in writing anything that to the best of my knowledge is not true. Errors are, of course possible, and my own.

The first two sections of this book are meant to give readers of any background an overall understanding of where we are in physics, logic and mathematics.

The section on electricity and magnetism seeks to give a general audience an understanding of electricity including inductance, how electric waves propagate and how magnetism fits in. I have delved more deeply into hydraulic analogies to electricity than, to my knowledge, has been done before. They aid understanding without the need of mathematics and extensive reference to mathematical principles which the reader may have never been exposed to or may not remember in depth.

The chapter entitled "Unified Theory of Gravity and Inertia" discusses Dr. Amitabha Ghosh's reasoning on the origin of inertia—it springs from the chapter entitled "Where to from Here and Now in Space and Time." Other areas discuss a few details in overviews that may be of interest to some readers including digital logic nd computers.

In the first edition I specifically talked about the term *inductance* as self-inductance, more properly, *self-induction*, by electric charge rather than the magnetic field induction generally given as the cause of inductance. On balance, I would now prefer to call a moving Coulomb force field the inducing cause.

This *Enhanced Magnetism Edition* also corrects or attempts to better explain the detailed words or phrases throughout the first edition.

In the other cases, this edition continues to summarize and extend the hydraulic analogies to electrical phenomena as the first edition did. It continues to frequently use fluid flowing through a hose with a flexible exterior within an enveloping fluid as the basic analogy to electric current flowing through a wire with its accompanying field stressing surrounding empty space.

Using this book

You can read this book like any other from front to back, perhaps picking up insights that you had not considered before, or you can choose to read only those portions of interest to you in which you are not already well versed. This book is focused on understanding, *not* building proficiency for use in any area.

The information I offer in this printed edition is what I consider essential to your reasonable understanding. I have added only what struck me at the time to be of some additional interest. I barely *dip my toes* into enormous amounts of information that is available on a particular subject. That can be found in books, courses or on the World Wide Web.

Optional Web links

If you can gain access to the World Wide Web, Google.com provides the capability for you to search into the largest library ever known. It provides extensive readily accessible supplemental information frequently including graphics, photographs and even courses tailored for a wide range of reader expertise.

I highly recommend its use for finding supplemental information that might be interest to you on a topic. I have found Wikipedia accurate as to information on physics. Of course, information for *any* source should be weighed against prior information that you have found reliable.

For your convenience I have frequently provided links that you can use to read augmenting material on the Web. If you have web access, go to giac2002.wordpress.com and click the heading at the top of the page entitled BOOK LINKS. On the page that appears titled: "Links for *Understanding Effects Across Space Enhanced Magnetism Edition*," hold the CTRL key down while simultaneously clicking the "click here" phrase associated with the particular link number that is associated with the particular Web article you found of interest in the book.

When you are through reading associated Web material, you will need to click the x at the top of your browser screen to the left of its address bar and return to where you were in the book text.

Unfortunately Internet Websites may change or disappear or your computer software may limit your connection capability. Specifically, note that reading a PDF file requires that you have its free Adobe Acrobat Reader software installed. I have checked and found that all the links provided in this book work with my computer.

Chapter 1
Understanding the Foundations

I believe we don't really know anything for certain although we may have strong beliefs.

Hallucinations or illusions aside, we are inclined to believe our own observations. But practically all of our beliefs are based on what other people tell us or write. These are believable to the extent that what they tell us appears to be logically consistent with our other observations, other things that we have accepted and our belief in the veracity and knowledge of the people relating their observations and beliefs.

Most of us tend to discount what people tell us if they can somehow benefit if we accept what they tell us, if they have told us something before that we have good reason to disbelieve or we do not accept their general veracity or knowledge of the subject of their assertions.

Another side of this story is that it does not make good sense to believe something is not true or real simply because it is not directly observable. For instance, the evidence for the existence of electrons is based on their effects. Although the effects of wind can be explained by the belief that air is made up of molecules, the belief that effects occur across what appears to be empty space [or more properly via a field] is supported by countless pieces of evidence alone.

General Take on Science, Logic and Mathematics

Early on in a field of interest, scientists collect observations. Then they attempt to see relationships among the observations. They abstract the observations into speculations that they cast in mathematical form to assure logical consistency—lack of internal contradiction. With mathematical manipulation they may be able to arrive at further logically consistent speculations.

Once a *speculation* is fully mathematically validated, it can become a *hypothesis* suitable for further test and observation that may lead to a *theory*. Scientists continually test any such theory for confirmation against experience. They no longer tend to call any of these theories *laws* except for those theories that were in the past called laws and continue to be called such simply because of custom.

Systematization of Logic and Mathematics Particularly in Science

Aristotle is generally credited as the first greatest systematizer of general logic specifically as to syllogisms and the principle of the excluded middle.

Euclid is generally credited as the first greatest systematizer of logic with mathematics as applied to the accumulated knowledge of plane geometry. He set the primary example for logically deducing provable results from a limited number of observations that are generally accepted as true or postulated truths. His method helped perpetuate the use of the *Occam's razor* principle that the simplest of competing theories is the one to be preferred.

Newton is similarly credited for formulating laws of motion and gravitation, Einstein for their even greater precision and extension in the cases of extreme velocities and in extremely strong gravitational fields. Many participated in the development of systematized quantum mechanical theories that successfully predict results of experiments at sub-atomic scales.

Mathematical systematization began with counting and arithmetic methods extending into manipulation of initially undefined quantities by algebraic methods. This was further extended to the rate of change mathematics of integral and differential calculus by Newton and Leibniz, theories of probability and many others including tensor analysis used by Einstein in his general theory of relativity.

Euclid's generalizations of the properties deducible for plane figures, plane geometry, have extended beyond the plane to any number of dimensions. Algebra and geometry were initially *married* by René Descartes into what turned out to be extremely enlightening and useful.

Logic was extended from the so-called *crisp* logic of the excluded middle to *fuzzy* logic that one way or another considered the *middle* in some probabilistic fashion. Crisp logic, as used in digital computers at least, basically seeks to just ensure consistency—not necessarily absolute truth. For instance, if a false premise is asserted in an *if* statement, the logic will still claim any conclusion is true.

Kurt Gödel used logic to show that there were logical limits of what it was possible to prove or disprove.

Mathematicians proceeded in such cases (or cases which might be provable but not yet) *by definition*.

The ultimate *decider* for any scientific theory or mathematical method is whether or not their results agree with observation and whether or not they are consistent—that is their constituent premises do not contradict each other. As Gödel showed, consistency itself can not be proven in any systematization of a complexity equivalent to or greater than that of ordinary arithmetic. Ultimately, a theory or mathematics is successful insofar as its results agree with observations and it is useful.

The degree to which mathematics applies to observed results is remarkable. In that mathematical methods must not be provably inconsistent, this makes sense. However, even the most beautiful mathematical theories may not *work* or be sufficiently useful. Hamilton's theory of quaternions is a prime example. It extended the manipulation of quantities in four dimensions in a way that beautifully paralleled proven methods in lesser dimensions, but was *ditched* for quite some time except for some concepts still used in vector analysis.

General Take on the Physical Sciences as Applied to Engineering

The aim in physical science as applied to engineering is to develop mathematical models so that calculations using the models will give sufficiently accurate predictions of the results that will be obtained from the usage of particulars in a real world application.

Newton's laws as to mechanical and gravitational effects are quote adequate for most applications. Maxwell's equations suffice for electromagnetic effects.

Einstein's theory of special relativity is required when speeds approaching that of light are involved and his theory of general relativity is required when very strong gravitational fields are involved or extremely precise measurements are required such as for the precession of the orbit of the planet Mercury [43 seconds of arc per century], the deflections of light in the gravitational field set up by earth's sun [1.75 seconds of arc], the displacements of spectral lines [1.00000212 greater wave length near the sun than near the earth] and corresponding fine corrections of time and global positioning satellite (GPS) measurements to account for the EGR [Einstein general theory of relativity] predicted slower time flow on earth than that on the satellite.

The so-called standard-model of quantum mechanics is required for accurate predictions and measurements in the subatomic realm—and those only as probabilities.

The standard-model does not yet account for gravitational effects—nor do gravitational theories account for standard-model effects although superstring theory holds out some—as yet incomplete—promise.

Classical Mechanics

Sir Isaac Newton was able to successfully mathematically describe the *laws* of motion for massive bodies and gravitation using mathematical tools such as algebra, Rene DesCarte's analytic geometry and the differential and integral calculus which he and Gottfried Leibniz contemporaneously developed.

The idea of *gravitation* was first proposed by Johannes Kepler as an attractive force between two objects. Christian Huygens proposed the major conceptual breakthrough that *force* (or *impetus*) is proportional to acceleration.

Newton used *mass* as a constant of proportionality. His *laws* of motion revolve around three basic premises—that force is proportional to the product of mass and acceleration, that objects not subjected to a force move at a constant velocity and that to every action there is an equal and opposite reaction.

He fit his gravitational model to previous observations and mathematical equations made by such luminaries as Galileo [based on inclined plane experiments on the rate of change of motion due to gravity] and Kepler [laws of planetary motion based on the precision astronomical observations of Tycho Brahe]. He developed an equation showing the force of gravity between masses was proportional to the product of the masses divided by the square of the distance separating them and originated the idea that planets and apples obey this same mathematical *law* of gravitation

Maxwell's Equations

Maxwell was able to successfully mathematically describe electromagnetic interactions modeled on experimental results by many, including Michael Faraday, and incorporating Faraday's concept of fields. By introducing a hypothetical current in dielectrics including empty space into his equations, his equations led to the prediction of the propagation of electromagnetic waves at a calculated velocity [the speed of light] based on basic electric and magnetic field constants. Experimental proofs of these predictions started with Hertz a few decades later. Unlike Newton's laws, they required no further refinement with Einstein's theory of special relativity

Einstein's Theory of Special Relativity [ESR]

Once you accept Einstein's rather strange premise that no signals or objects move faster than the speed of light along with his extension of Galileo's premise that it is impossible to detect motion within a frame that is moving at a constant velocity relative to other objects so as to include measurements of the speed of light and consider the consequences, it becomes evident that the length of objects and time intervals between events appear different to different observers moving at a constant velocity [constant speed in a constant direction] with respect to each other. Perhaps, Einstein's most important insight was that measurement is the fundamental method for the acquisition and application of physical knowledge to prediction—that is, measurement is the only precisely true basis we have for assessing physical reality.

Although, given his premises, only simple high school algebra is necessary to arrive at Einstein's basic equations of special relativity precisely defining the amount of time and length shrinkage, the concepts that they precipitate are mind-boggling. One is the conclusion that the *measured* age of twins differs markedly as their relative velocities approach that of light. [In his theory of general relativity, he establishes that the acceleration and deceleration necessary for the twins to return again to the same *inertial frame* would result in the same absolute age difference as the measured age difference when they were in relative motion.]

Once it is accepted that length and time measurements are dependent on relative velocity, it becomes apparent that the measurement of some other physical quantities will also be dependent on relative velocities. Specifically, the theory shows that the apparent masses of bodies increase with their increased relative velocity. The calculated mass increases have been experimentally verified with measurements made on particles moving at relative velocities approaching that of light—such as within cyclotrons.

Further derivations on the same bases resulted in the famous $E=MC^2$ equation which shows the enormous amount of energy available in matter—that is, the energy available in matter equals its mass times the speed of light squared. This has been corroborated by the enormous energy released in nuclear bombs and explains how earth's sun is still radiating energy billions of years beyond that possible by the ordinary, non-nuclear burning of fuel.

The basic equations relating the observation and thus measurement of length and time varying with the relative speed of observers are identical with the previous Lorentz-Fitzgerald equations which were descriptive of observations with no other explanation. Einstein derived them from two basic apparent truisms: the measurement of the speed of light is identical to all observers regardless of their relative speed and the more general extension of the principle that mechanical observations are the same as to electromagnetic effects regardless of the velocity of the reference frame.

The Lorentz-Fitzgerald outlook was that lengths themselves actually shortened at velocities approaching that of light not just their observed length. Once one accepts that nothing can exceed the speed of light and so no part of an object can, it is reasonable that a body will actually shorten as that speed is approached. ESR shows that it is the *measured length* that shortens.

Einstein's viewpoint makes practical sense as scientific applications deal with observation and so measurements. An example is the measurement of the increased mass of particles moving near the speed of light in an accelerator is as *real* as it gets even though it is *really* due to the relative speed of the observers/measurers.

Einstein's Theory of General Relativity [EGR]

Einstein, understandably, wanted to determine what the results would be if the observers were moving at other than a constant velocity with respect to each other. In doing so, he observed that gravity between objects was an effect indistinguishable from a constant relative acceleration [that is, the change in the relative speed or direction of motion, i.e. relative velocity, between objects]. And he hypothesized that gravity was caused by a distortion of space as well as a distortion of time in the presence of mass/energy—an effect that was caused by the mass/energy of objects rather than a force as Newton hypothesized. Of course, in both cases, the interaction of gravity and mass is an observational given.

Objects within this space/time gravitational field would move along *geodesics* determined by that field rather than along the Euclidean straight lines defined by Newton's mechanics.

Einstein then set out on an 11 year long search for an equation or equations that would most exactly describe the curvature of space and time and the geodesics caused by the mass/energy that corresponded to gravity. The mathematician Hermann Minkowski had shown that Einstein's special theory of relativity could be described as a vector in space-time. The projections on the time and space axes of the vector representing the different time and length measurements made by observers moving at a constant velocity with respect to each other, whereas both observers would agree on the measurement of the vector itself—that is, the vector itself was an *invariant*.

It takes 4^4 or 256 equations with 256 unknowns to describe all the possible distortions of four dimensional space-time due to a gravitational field. Einstein's mathematician friend and former school-mate, Marcel Grossmann, pointed out that the mathematicians Gregorio Ricci-Curbastro and Tullio Levi-Civita had finished development of tensor calculus a little more than a decade before [c. 1900 AD] that would allow simultaneous manipulation of all these equations and the resulting tensor [a multiple-dimensional generalization of a two dimensional vector] would be an invariant measured the same by all observers.

With use of the techniques of tensor calculus and more standard mathematics, the 256 equations reduced to only three upon introducing terms representing gravitational fields. Einstein and Grossmann developed a four dimensional curvature tensor applicable to gravitational fields in a joint paper with other mathematicians *kibitzing*. Grossmann left to Einstein any physical interpretation of the tensor. It took Einstein another year or two to satisfy himself that their derivation was applicable to the real world with his final choice made to most closely correspond to the proven gravitational description by Newton.

He was delighted to find that his final choice also explained the 43 seconds of arc precession of the orbit of Mercury per century beyond that of previous calculations with Newton's theory based on known gravitational perturbations of the other planets.

Two expeditions were planned to test Einstein's resulting prediction of the deflection of light from a distant star observable when a light beam passed near to the sun in a total eclipse. Fortunately, weather caused a cancellation of the first, because Einstein discovered needed modifications in calculations which doubled his first prediction. His second prediction was different than what Newton's laws would predict even considering the apparent mass of photons which Einstein's special theory showed.

Publication of the second expedition's results being within accepted margins of error made Einstein famous world-wide. Einstein's equations still required him to use mathematical approximations.

He was surprised by Karl Schwarzchild obtaining an exact solution of his field equations within little more than a month after their publication—apparently based on a minor fine tuning of the physical situation abstraction, that is, by specifying a spherical, uncharged non-rotating mass. Schwarzchild's results were the same that Einstein had obtained with mathematical approximations.

Einstein's basic equations remain non-linear [due to the fact that gravitational fields represent energy and so must be included in the mass intrinsic to the calculations]. This non-linearity and complexity due to the multiple equations that must be satisfied make their solutions difficult. The simplicity of their underlying assumptions are what make them consistent with the philosophical *Occam razor* standard of elegance.

His theory of general relativity has successfully predicted the degree of shift in the spectral lines of elements as well as other time shift measurements in gravitational fields and provided the basis for otherwise needed global positioning satellite distance corrections.

In 2016, precision instrumentation led to observation of gravitational waves in space as theorized by Einstein in his general theory of relativity.

Any disproof or bettering of his theory of general relativity by others in the nearly century since its inception would *make* their reputation. Ever more precise measurements have accorded with his theory and attempts to better it have not received general acceptance.

Nevertheless, he did not receive his Nobel prize for either theory of relativity as they were not yet then sufficiently verified to convince everyone in the scientific community. Instead, he was given the prize for his paper in the same year as the one he wrote on special relativity, this time an original paper on what became quantum mechanics.

The *Standard* Theory of **Quantum Mechanics**

Max Planck had proposed what Planck considered a *fudge factor* to explain experimental anomalies in classical electromagnetic theory predictions. Perhaps the most understandable is its eventual successful resolution of what has been named the *ultraviolet catastrophe*.

In classical theory an ideal black body in thermal equilibrium would radiate power in proportion to the frequency squared. Thus the total radiated power would be unlimited at higher and higher frequencies. This does not jibe with observations.

Planck found that that theoretical fallacy could be resolved if one were to stipulate that energy could not take on any continuous value but could only be emitted in discrete packets proportional to frequency. The precise mathematical relationship was dubbed: *Planck's constant*.

Einstein invoked Planck's constant in theorizing that light was quantized in what he called photons in its discreet reactions with photoelectric material rather than what would be expected with a purely continuous electromagnetic wave.

This paper by Einstein was the initial paper leading to other successful theories invoking Planck's constant which among other discoveries led to the generalized standard theory of quantum mechanics—although Einstein rejected the overall quantum theory that developed on philosophical grounds because it did not jibe with his belief that a true theory would yield precise fixed rather than simply probabilistic predictions for observed results.

Einstein respectfully but unsuccessfully argued with Neils Bohr for decades on the subject. However, the *standard theory* predictions—made by numerous theoreticians and experimenters—has remained astoundingly accurate. Richard Feynman in his popular book *QED* [quantum electrodynamics] described a prediction: "If you were to measure the distance from Los Angeles to New York...it would be exact to the thickness of a human hair...There are other things in quantum electrodynamics that have been measured with comparable accuracy...These numbers are meant to intimidate you into believing that the theory is not too far off!"

Unfortunately the quantum theory standard model is not generally believed to meld with any established theory of gravity.

Superstrings and Physical Models

Eleven dimensional superstring theories have held out some hope to successfully combine Einstein's theory of relativity with the standard model but they have not yet, apparently, survived all theoretical verifications.

The mathematical theory has been described as a physical model involving open and closed strings at the order of Planck sizes. Such sizes are considerably below anything that can possibly be, or certainly currently, observed. Different particles are theorized to be manifestations of different vibrations of superstrings or generalizations called *branes*.

One might question whether anything unobservable is legitimate as a physical theory, but molecules, atoms and subatomic particles were originally unobservable although they became accepted as real objects because of the observable results that their theory predicted.

Electric Charge, Magnetism and Electromagnetic Fields

To me, and perhaps to many people generally, actions that are observed to occur across the vacuum of empty space are most interesting because they seem to defy what we intuitively feel make sense. Einstein has de-mystified gravity somewhat if one accepts that even empty space can be deformed by mass in such a way that other masses react.

The observed facts that like electric charges repel and unlike charges attract are similar to gravity in that they are just that—observed facts for which there are no known exceptions.

Einstein was apparently unable to encompass the interactions between electrically charged and magnetically coupled bodies in equations such as those he used ascribing gravity to space-and-time distortion. However Einstein's theory of special relativity does explain the previously unexplainable magnetic interactions due to relative motion between charges.

For instance, the density of charges in a current carrying wire with respect to an external charged particle moving parallel to the wire will be the greatest, due to special relativity, when the relative velocity of the charges in the internal wire and the velocity of external charge is the greatest. If the external charge is an electron moving in the same direction as electrons flowing within the wire, the greatest relative velocity will be with respect to the protons in the wire so the negatively charged external electron will be attracted to the more densely distributed positively charged protons in the wire. This attraction is called *magnetic*. If the external electron is moving in the opposite direction, the repulsion—due to the motion—is also called *magnetic* even though both effects arise from the force between charges.

It turns out that *all* magnetic effects can be ascribed to the relative motion of charges if one accepts notions [that seem to be effectively true] that matter consists of protons and relatively freely moving electrons that can spin, orbit protons and move relative to each other.

World Wide Web Lectures

You may choose to visit giac2002.wordpress.com and under the BOOK LINKS heading, put your mouse pointer on the word "here" associated with link 1, hold down the CTRL key and click to view outstanding YouTube lectures by Professor Michael J. Ruiz of the University of North Carolina at Asheville on electromagnetism and associated mathematics.

You may also choose to visit giac2002.wordpress.com and under the BOOK LINKS heading, put your mouse pointer on the word "here" associated with link 2, hold down the CTRL key and click to view a video lecture that graphically exhibits and shows the mathematical derivation of the magnetic force arising from special relativity in combination with Coulomb's laws of like charge repulsion and unlike charge attraction.

Aside: Professor Ruiz mentions the textbook by Edward Mills Purcell: *Electricity and Magnetism, Berkeley Physics Course*—Volume 2. Purcell was a 1952 Nobel Prize winner with Felix Bloch "for their development of new methods for nuclear magnetic precision measurements and discoveries in connection therewith."

Ruiz used the Purcell book as a source for his slightly modified approach to the derivation of the magnetic field from special relativity and Coulomb's law: *The usual approach is to take both positive and negative charges moving opposite each other and exploit the symmetry. But the realistic case is to have the electrons move and not the positive ionic cores of the atoms in the metal.* Ruiz's approach is what I have further and hopefully accurately clarified in my non-mathematical description.

Summarizing the video lectures above on this approach—Wires have both protons and electrons as do all materials. And when electrons are flowing through one of two parallel wires, the protons in the other wire are moving at the speed of the electrons. So by relativistic contraction, their effective density from the perspective of the electrons increases and so their cumulative effective charge from the perspective of the electrons is positive. This then results in a so-called magnetic attraction between the two wires.

The magnetic effects of electric currents had been determined experimentally years before and equations detailed that have appeared to be completely adequate before the fundamental reason tied to special relativity was generally known and accepted. When I first learned of the proposed reason, it seemed a reach, at least to me, that the electron drift in conductors could cause the known magnetic results. But when Purcell who received his Nobel laureate for magnetic pursuits wrote a textbook using this explanation, it pretty much closes the case.

Conclusions

There are two basic unexplainable characteristics of matter, mass and charge. The basic properties of mass are gravity and inertia. Gravity is a given whether its effects on other masses are explained by force across empty space or local curvatures of the space-time continuum in accordance with Einstein's theory of general relativity that can be summarized this way: *Mass-energy distorts space-time and the distortion of space-time determines the way that mass-energy moves or propagates.* Inertia has been explained as the effect of gravity of distant matter and has been quantified by Amitabha Ghosh with a modified theory of gravity that has not been generally accepted or ruled out.

The basic effects of charge over space [or, more properly via a Coulomb force field] that like charges attract, unlike charges repel and charges in motion cause magnetic effects.

Charge interactions might conceivably be explained as distortions of a space time similar to those for gravitational interactions, but neither Einstein nor anyone else has succeeded in doing so—even though moving Coulomb force fields along with the Einstein theory of special relativity explain magnetic effects. Because only charges are affected by charges, it would seem that one might have to suppose that charges distort a special space-time that affects only charges?

Mathematics has and continues to be important in practically applying and arriving at a quantitative understanding of physical effects—geometry, algebra, calculus for Newton's theories, vector analysis for Maxwell, tensor analysis for Einstein's theory of general relativity to name a few. Einstein was unimpressed initially, but came to depend on mathematics to lead the way in general relativity. Nowadays, arguably the most advanced theoretical approach to seek a grand unified theory including relativity and the standard model of quantum theory is through mathematics.

Faraday's field concept underlies the physical understanding of, at least, electronics, magnetism, and general relativity.

Chapter 2
Mathematics

Introduction

Mathematics provides tools for detailed analyses and practical usage. The coverage here provides an overview of various mathematics to generally describe them, their place and their basic concepts for those who have not been exposed to them, may have forgotten them or concentrated on their usage with little regard for basic principles.

Practical usage requires further study to learn how to set up equations for particular cases and solve them by personal mathematical manipulation or use of calculators and computers. Use of the last is much less prone to detailed mistakes.

Whereas theories in the physical science are plausible generalizations based on observations and *proven* to the extent that exceptions to predictions from those generalizations are not observed, mathematical principles are generalizations that are *proven* to the extent that they do not lead to logical contradiction.

Mathematics is characterized by the introduction and systematic use of names and symbols and their relationships, logical consistency and the accumulation of proven [internally consistent] theorems for further usage. Only a few mathematical symbols on a page may be adequate to describe a wealth of results with physical phenomena.

After relationships are exhaustively examined and found devoid of contradictions in, sometimes, generations of use by hordes of people, their further use can be reasonably relied upon without, every time, going through all the logical steps to *prove* them.

As mathematics and physics evolved, use of generally agreed upon standardized names and symbols to represent combinations of basic concepts and operations was found to be efficient and immensely useful in removing the ambiguities in natural language. Using mathematical methods can be likened to turning a crank to arrive at a solution or solutions once the basic mathematical relationships are set up.

Our belief in physical science results and the correspondence of mathematical predictions to the actual physical results can not always be personally or practically verified. We often accept or, practically, have to accept what we are told has been verified or is generally accepted, e.g. the *little g* gravitational constant on a particular distant planet [the free fall acceleration at a planet's surface as opposed to the *big G* universal gravitational constant essentially established by Henry Cavendish 71 years after Newton's death]. We are sometimes forced to accept the statements of those we believe are experts or the consensus of their opinions.

Mathematical theorems can, however, be generally verified using our personal ingrained sense of logic and earlier developed theorems using established logical principles that have held for extended lengths of time. Gödel however has proven that a given theorem may not be provable in principle, not even considering one which no one has succeeded in proving. And those totally consistent theorems with no internal logical contradictions may not be applicable to the real world. Such applicability must be separately shown.

A problem with mathematics and computers is their users' very lack of involvement in the details beyond setting up the equations or the programs. This may impede intuitive understanding and intuitive development. Faraday, the foremost pioneer in electrical and magnetic experimentation as well as developer of the *field* concept that Maxwell formalized was put off by what he considered the obscurity of Maxwell's mathematical development. And Einstein was inclined to denigrate mathematics until he became converted with the power of tensor calculus.

However without the shortcuts to detailed reasoning offered by mathematics, Maxwell might never have deduced that electromagnetic waves of any type move with the speed of light, despite Faraday's previous intuition as to possible wave propagation. And Einstein may well have not been able to derive the extremely accurate corrections of his general relativity to Newton's law of gravity that are essential in many applications even though negligible in most.

Mathematics and physics have necessarily evolved languages of their own to avoid having to discuss everything over and over again from basic principles.

The venerable Chemical Rubber Company [CRC] *Standard Mathematical Tables* serve as a relatively terse virtual mathematical encyclopedia and dictionary. In addition to its compilation of many proven useful mathematical tables, including logarithms, and instructions for their use, it compiles conversion factors, miscellaneous constants, theorems, mathematical definitions, formulae, mathematical symbols and abbreviations.

Depending on your background, feel free to skip over the subsections of this chapter with which you are already fully familiar. They are meant primarily to remove the mystery of a particular mathematical area for those without such background or maybe to augment understanding. I don't believe any are absolutely essential to understanding the other chapters.

Counting, Numbers and Arithmetic Processes

The concept of number and numerical symbology underlies all mathematics. I've included this section with its everyday concepts simply because I realized there were some underlying ideas or relationships that I had not thought through, or maybe long forgot, and I believe there may be others interested and in the same boat.

The first historical realization was that numbers and counting were generally applicable—not just to sets of particular type of objects—like, say, sticks in some languages.

Numerals are, of course, names or symbols used to count. The most commonly used ten basic numerals are generally believed to have resulted from use of the ten fingers (digits), when the thumbs from both hands are included, to count. Any base including the base ten can be used.

[Asides: The binary two digit system has formed the computational basis of modern computers. Babylonians used 60 as the basis of their mathematics (so 60 second and 60 minute divisions of arcs and time). The last allows many more even divisions than the decimal system, but obviously requires many more basic numeral names and symbols.]

Counting the *total* or *sum* of objects or symbols in a group or groups of objects or symbols is the most primitive form of addition. You can then *take away* or *subtract* a counted number of objects and count the *difference* remaining or repeatedly subtract a counted number of those objects a counted number of times to *divide* the original groups to find the *remainder*. Similarly, you can count a number of objects and add them a number of *times*, or *multiply* them to form a *product*. The products of the *factors*—the original number of objects and the number of times they are to be multiplied—can be arranged in a *multiplication table* that can be referenced or memorized to determine their product without actually counting them.

You can *take away* or *subtract* a counted number of objects to arrive at a *difference*. Or you can *divide* by taking away groups of objects repeatedly in which each group contains the same number of objects, the *divisor*, until that number of objects no longer remains available. Any number remaining is the *remainder*.

After choosing a symbol to represent an object, say a vertical line, you can tally them by grouping the symbols, say by drawing horizontal lines through groups of five and then come to a total by counting the groups and multiplying by the number of groups.

This idea became mechanized with counting frames or forms of abaci. A simple form of a counting frame or abacus could be constructed with a set of vertically mounted wires having nine slide-able beads on each wire that are initially on the bottom of all the wires.

Figure 1-1 Idealized counting frame with nine slidable beads each on a separate wire

The Japanese soroban style counting frame is more representative of counting frames used commercially. In addition to wires holding as many as four beads each, it includes a wire with only one bead. In counting, just after one of the wires with four beads has filled, on the next count, the bead on the wire containing only one bead is moved to its non-zero position while the wire that just previously held four beads is returned to its zero position. The results are just like typical tallying marks in which the number of beads in the non-zero single bead wire position represents the number of fives tallied.

To count one bead at a time, you move the lowest positioned bead in the rightmost wire up one-at-a-time until that wire is filled with nine beads. Then, a bead is slid up one position on the wire immediately to the left of the first used wire and all the remaining beads on the first used wire are moved to its bottom. By continuing in the same fashion you can determine the count of any number of objects up to the amount when all the paralleled wires are fully occupied with beads.

The resulting configuration is traditionally symbolized in the West with numerals placed in a series to form a number, each position representing a wire left to right in the simple abacus just described and the number of beads on each wire by the value of the numeral in each position. The numerals in a given place represent ten times the value of those in the adjacent place to its right unless all the beads symbolized in a row are in their bottom-most position that is symbolized with a placeholder, traditionally the numeral zero.

A numeral other than zero in each position then represents one-tenth that of the value of the same numeral in the position just to its left. When we place a decimal point to the left of a position, the numerals in that position and those in positions to the right represent successively smaller fractional numbers, i.e. one tenth, one hundredth, etc.

We can obviously move beads or manipulate the corresponding symbols to perform all the arithmetic operations that we could by counting groups of objects. Many algorithms or various methods for determining other arithmetic processes via an abacus or written representations have been developed.

Various forms of counting frames are used. For instance, rather than cash registers, the Japanese have used their soroban counting frames in commerce for generations.

Figure 1-2 Japanese soroban counting frame

Photo: Jordyn Gum

In the sorobon, the upper level of beads represents zero or five units depending on their placement in the upper or lower positions as shown in figure 1-2.

The general concept of fractional numbers appears often difficult for school children to grasp. The top term or *numerator* is merely the number of parts chosen from those that the lower term or *denominator* divides something into. The bar separating the numerator and the denominator is effectively just another symbol for a division that does not result in a remainder.

Fractional notation is also used to express the ratio of two quantities. So fractional numbers are also *rational* numbers, and quantities that can not be expressed as fractional or whole numbers are called *irrational* numbers.

Georg Cantor gave easily followed and generally accepted elegant arguments in real analysis regarding infinite numbers. They conclude that there are an infinite number of rational numbers, an infinite number of irrational numbers and an infinite number of either between any two real numbers. The set of all *real* numbers represents <u>all</u> the numbers on a line—that is *all* the decimal numbers. Every point on a line can be expressed as a real or decimal number.

Symbols and Language

Symbols for counting operations include **+** for addition [*plus*], **x** or an elevated dot for multiplication [*times*], a superscripted number for exponentiation [that is, the number of times the basic number is to be taken times itself] e.g. N^2 for a basic generalized number here symbolized by **N** that is called *taken to* [or *raised to*] *the second power* or *squared* and their respective inverses: **-** for subtraction [*minus*], **÷** or **/** for division [*divide by*], and negative exponentiation for the reciprocal of a number taken times itself, e.g. N^{-3} for a generalized basic number here symbolized by **N** *taken to the third power* or *cubed*—and then divided into one.

The expression *squared* for a number taken to the second power is representative of the fact that a length taken times itself is the area of a square with sides equal to that length. Similarly, the expression *cubed* for a number taken to the third power is representative of the fact that length taken times itself three times is the area of a cube with sides equal to that length.

The square root of a number is the number that when squared will equal the original number. The square root operation has been traditionally symbolized by $\sqrt{}$. That is, the square root of a number **N** or the operation to find the square root of **N** can be symbolized as $\sqrt{N}$. It could also be symbolized as $N^{1/2}$. The nth root of a number **N** can similarly be symbolized as $N^{1/n}$, meaning the number which taken n times itself would equal **N**.

[Aside: Exponents, other than just being a compact way of showing the operation of multiplying a number by itself the number of times shown in the exponent, have useful properties—for instance, the product of two base numbers in exponential form is the base number raised to the sum of their exponents. For instance, $N^{1/2} \times N^{1/2} = N^1 = N$. That shows that the product of the square roots of a number is the number, as it should be.]

Any number to the zero power is defined as 1.

Logarithms are the exponent to which a base such as 10 must be raised to produce a given number. For instance, because 10 to the third power is 1000, the logarithm of 1000 is 3 to the base 10. One important use of the logarithms is that the log of a product is equal to the sum of the logarithms of the factors.

Because of such conveniences, extensive tables of logarithms are available. And in the past slide rules laid out on sliding logarithmic scales were made so that sliding one inscribed number to align with another was tantamount to adding their lengths so their product could be read off. They were commonly used for calculations by engineers before the advent of electronic calculators.

Other calculations in addition to multiplication can be simplified with logarithms. For instance, division can be simplified by simply subtracting the logarithms of two numbers. For instance 10 to the sixth power divided by 10 to the third power, that is 10,000,000 divided by 10,000 is equal to the difference of their exponents [their logarithms] applied to the base 10—that is, 10 to the second power which is 100].

Euclid's Plane Geometry

With only 5 axioms (alleged self-evident truths) and 5 postulates, Euclid recorded techniques for proving almost limitless relationships among geometric figures on a plane. His approach of proving so much using only a limited group of observations or perceived truths has proven basic to all fields of mathematics and physical science.

Before Euclid's geometry, science and mathematics consisted of collecting information and making inferences based on that information or inductive reasoning. Euclid's methods set the subsequent pattern for deductive reasoning based on the assumed truth of generalities.

Euclid's axioms
1. Things which are equal to the same thing are also equal to one another.
2. If equals be added to equals, the wholes are equal.
3. If equals be subtracted from equals, the remainders are equal.
4. Things which coincide with one another are equal to one another.
5. The whole is greater than the parts.

Euclid's postulates
1. One can draw a straight line from any point to any point.
2. One can produce a finite straight line continuously in a straight line.
3. One can describe a circle with any center and distance.
4. All right angles are equal to one another.
5. Through a given point P not on a line L, there is one and only one line in the plane of P and L which does not meet L [as re-stated by John Playfair in 1795].

Other than exhibiting the sheer explanatory power and reasoning for many proven results from a few assumptions, some knowledge of plane geometry can prove useful in other mathematics. The Pythagorean theorem stating that the length of the third side of a planar right triangle [one in which one of its angles is 90 degrees] is equal to the square root of the sum of the squares of the lengths of the other two sides was of such practical significance that it was known in antiquity perhaps even before Pythagoras [570-495 B.C.]. Proofs preceded those of Euclid. The important fact that the sum of the angles of any planar triangle is 180 degrees is easily proven by Euclid's methods and is an essential fact for trigonometry.

Trigonometry

Trigonometry consists of generalized methods for determining unknown sides or angles of plane triangles that are easily extended to other plane figures bounded by straight lines. These are facilitated by the naming of the ratios of corresponding sides of right triangles—shown by the methods of plane geometry to always be the same for triangles with the same second [other than the *right* 90 degree angle] angle.

Tables of these ratios are used extensively in scientific and engineering applications. In surveying, for one practical instance, any plane parcel bounded by straight lines can be described by a set of contiguous triangles.

The traditional names for the ratios are: sine for the ratio of the side opposite of the named angle over the hypotenuse (the side opposite the right angle), cosine for the ratio of the side adjacent to the named angle over the hypotenuse and tangent for the ratio of the side opposite to the side adjacent. The traditional names for the *flipped* or inverse ratios are secant, cosecant and cotangent.

The term tangent can also be expressed as *rise-over-run*. The word *tangent* comes from a Latin term meaning to touch. In mathematics it also describes a line touching a curve in only a single point.

Algebra

Basic algebra consists of generalized methods for calculating the value of unknown quantities using the substitution of symbols called variables rather than specific numbers in arithmetic equations. Algebraic techniques, next to arithmetic, are the most basic and the most used mathematics in science and engineering.

Algebra may not be as glamorous as calculus, that is nearly essential in deriving and understanding the mathematics behind many physical principles, but understanding and using calculus would be impossible without knowledge of algebra.

If mathematics is the *queen of science*, certainly physical science, algebra is the *queen of mathematics*.

Algebra typically uses a particular letter of the alphabet for a particular type of number, e.g. a might represent the number of apples, o might represent the number of oranges, etc. That is, to set up an algebraic problem to solve, you use a symbol, such as a as a shorthand for the entire phrase: *number of apples* and then set up known relationships. For instance, you might know that the price of an order is the number of apples times the price of one apple plus the price of one orange times the number of oranges. If the price of an apple is 50 cents and the price of an orange is 60 cents, and the total price of the order in cents is symbolized by T, you would write: $T = 50a + 60o$. Going one step further, you might symbolize the price of one apple as A and the price of one orange as O and write $T = Aa + Oo$.

The basic idea in algebra is that the result of the identical arithmetic operations performed on *both* sides of an equality does not change the overall equality relationship. That is, except for extending addition and subtraction to other arithmetical operations, the basic idea reflects the first three axioms of Euclid's plane geometry. This idea is the same as what is called transposition in algebra—that is, if you transpose by moving any term of an equality to the other side of an equal sign, the term will take on the inverse operation [e.g. +1 on the left side when moved to the right side will become -1]. It also allows the removal of the terms that are involved with the same variable in two equations by rearranging such terms so they are aligned and then subtracting one equation from the other. Continuing in this manner for multiple equations allows all the variables but one to be eliminated until only one equation remains to be solved—as long as there are as many independent equations as variables.

As in any other branch of mathematics, many general relationships have been derived and may be useful—from memory or from tables in books such as the Chemical Rubber Company book listed in the References section. The so-called quadratic equation, not described here, may be one of the most generally useful in algebra and so possibly worth-while to memorize.

Learning how to use algebra in practical situations or *word problems* can be one of the most difficult that a student encounters in high school. However, in addition to its practicality, one who perseveres can be amazed and gratified when he or she puts solutions back into the original problem and finds them correct.

The general procedure is to define what your symbol letters are to stand for, then use them to express known information in equalities followed by manipulating the equalities algebraically in such a fashion as to express the value of a variable in an equality with a constant number—the *solution*.

Feel free to skip the equations that follow if you already have a background in algebra or you feel that further explanation is unnecessary for understanding.

As an example for those who have never had a course in algebra: A problem could be to determine if you have fifty pounds of a cement-and-sand mixture that is 40% sand, how many pounds of salt must you add to obtain a mixture of 60% sand?

Algebraic solution:

Let C = pounds of cement
S = pounds of salt
a = additional pounds of salt to be added
$C + S = 50, S/(C + S) = .4, S/50 = .4, S = 20, C = 30$
$(20 + a)/(50 + a) = .6, 20 + a = 30 + .6a, 0.4a = 10, a = 10/.4 = 25$

Check:

$20 + 25 = 45$ [total pounds of salt in the final mixture]
$50 + 25 = 75$ [total pounds of final mixture]
$45 / 75 = .6$ [fractional amount of salt in the final mixture]

It takes a lot of words to describe the example procedure, but here it is. After first expressing the words given in the example with two equations using the symbols described, the steps above are separated by commas: we substituted the known weight of the mixture for the symbols for the sum of the individual weights and multiplied both sides of the resultant equation by that sum to obtain the number of pounds of salt in the original mixture. The number of pounds of cement follows directly.

We then added the symbol for the as yet unknown additional pounds of salt to the initial pounds of salt and divided by the sum of the initial total weight of the mixture plus the added salt to determine the fractional amount of salt in the final mixture which we set to equal 0.6, the fractional amount corresponding to the desired 60% in the final mixture. Then we multiplied both sides of the equation by the divisor in the first term to obtain a new equation $0.4a = 10$ which we can express in terms of $1a$ by dividing both sides by 0.4 to arrive at $a = 25$.

In the check, we summed the added salt of 25 pounds to the original salt weight and the original combined weight, took their ratio and found that it was indeed that desired.

The procedure above and the illustrative example provide solutions in many, perhaps the majority of cases, involving only two initial unknowns. In cases involving more than two unknowns, once one of the unknowns is eliminated by algebraic manipulation and subtractions, the same procedure can be used on the remaining equations until only one equation remains involving only one unknown.

In the event, that any result is more involved than an equation simply showing the unknown equaling a number, there may be a previously proven general solution—for example, the quadratic equation which is not detailed here although you can google it.

A standard reference book containing tables and equations for the many branches of mathematics, logarithms, standard mathematical terms and symbols is the Chemical Rubber Company's *Standard Mathematical Tables*.

Coordinate Systems, Mappings and Analytical Geometry

The idea of coordinate systems although almost boringly apparent underlies much of more advanced mathematics. They provide a geometric mapping and thus an intuitive grasp of the relationships among mathematical variables in the three dimensions that we can actually observe. Mathematically, there is no limit to the number of dimensions that can be used even though geometry beyond three dimensions in not generally intuitive.

The Cartesian system (decisively developed by Rene Descartes [1596-1650]) mimics planar geographic north-south and east-west mapping with respect to the equator and the prime meridian except mathematicians name the corresponding geometric map directions as positive or negative x and y values with respect to arbitrarily placed horizontal axis and vertical axes rather than *north* or *south* or *east* or *west*.

Figure 2-1 Representation of complex number plane

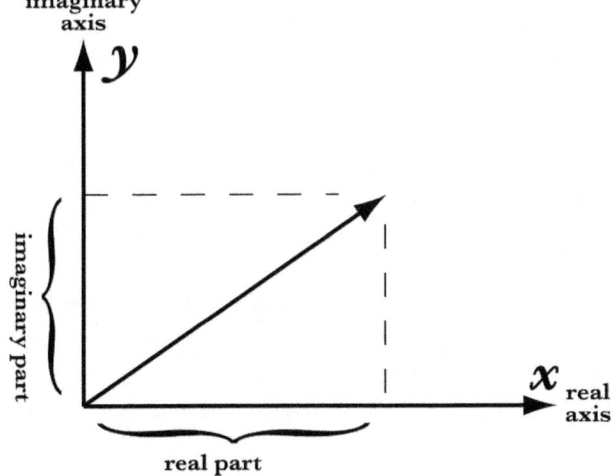

Extending this concept to three dimensions, the letter z is typically used to represent a dimension perpendicular to the other two x and y axes.

Figure 2-2 Representation of a three-dimensional rectangular coordinate system

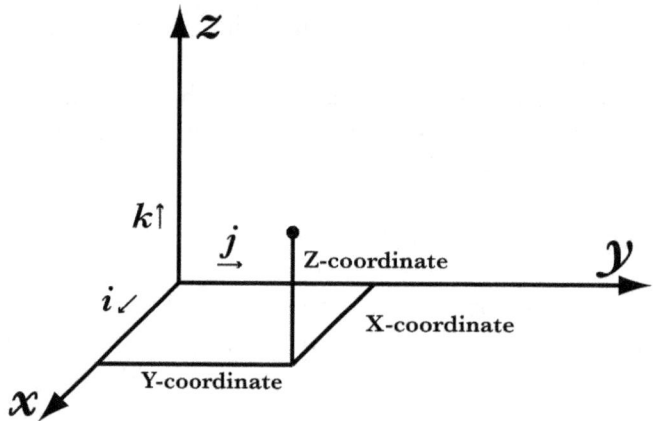

The notations for the axes of these or other coordinate systems can be and are frequently generalized to such as x_1, x_2 etc. to any arbitrary x_n.

Other common coordinate systems use the fact that a point can be used to precisely locate points with respect to the location of the center of a circle and the angular position on the circumference of such a circle of given radius. On a plane, such a coordinate system is called polar or circular. In three-space, a coordinate system consisting of two circles at right angles to each other and the distance to the origin is called spherical. Angular positions are denoted by various names, perhaps the most intuitively apparent are azimuth and elevation. A coordinate system consisting of a circle at right angles to a line with given height is called cylindrical. The coordinates are the distance to the axis of the cylinder, height and azimuth angle.

The *locus* [mapping] of possible values of variables of any algebraic equation can, of course, be plotted in any such coordinate system. Such loci, even when they are or contain straight lines are called *curves* or *functions*. Similarly, any such curve can be represented by an equation.

Analytical geometry can be described as the systemized methodology to derive and accumulate the general equations of various curves. Perhaps surprisingly, common geometrical curves, specifically conic sections (circles, ellipses, parabolas and hyperbolas) and straight lines correspond to straightforward algebraic equations.

The accumulated concepts, general relationships and equations results are used in other branches of mathematics and specific derived results in mathematical applications to physics.

Calculus Terms

If you have had previous exposure to the calculus, feel free to skip this section entirely. But one of the reasons that people's eyes tend to gloss over when anything is discussed in mathematics in general and calculus in particular is that the terms used are unknown or inadequately understood.

The great contribution of analytic geometry to understanding was to show how equations appear geometrically. We tend to *see* things more readily if they are rendered in graphical form.

Much is made of the term *function* in calculus and advanced mathematics. In basic calculus when the horizontal axis is designated in terms of a variable designated by x and the vertical axis is designated in terms of a variable designated by y, it is a *mapping* of the variables on the xy plane. It may, loosely be called the curve although the resultant *curve* may actually be a line. A function of a complex variable refers to a mapping from one complex plane to another. A complex plane is one in which real numbers are represented by points on one coordinate axis, typically the horizontal axis, and imaginary numbers on another and complex numbers are represented by their mapping on the plane. In more specific terms, a function is a mapping of one set of variables to another set—in basic calculus, so that one mapped value is related to one mapping value.

The term function undoubtedly arose from the fact that the value of a *dependent variable*, in this case y, is a function, in the most ordinary language, depending upon the *independent variable*, in this case x. That is, as x varies horizontally, the value of the function varies in the vertical direction correspondingly. Unfortunately, it is often said that y varies, but what is more precisely meant is that the function varies [in the y direction].

A function in this case, may be designated as $f(x)$ and spoken of as "f of x". In general, it might be designated as $g(w)$ when the independent variable runs along the horizontal axis and g is a different function. For instance, $f(x)$ might be represented as $y=10x^2$ and $g(w)$ as $z=7+w$ when z is the variable along the vertical axis.

The terms endemic to the differential calculus are, as you may have guessed, differential and also derivative. The term in differential in basic calculus refers to an infinitesimal distance that approaches zero as a limit along one axis or another.

In rectangular coordinates, the term derivative refers to the ratio of the differential distance in the vertical direction divided by the differential distance in the horizontal direction. Unfortunately, this derivative is often referred to when dealing with a function of x in x-y coordinates as dy/dx when it more precisely means $df(x)/dx$ for the function $f(x)$ at a particular point.

It turns out that such a derivative is the tangent to a curve at the point on the curve that the derivative refers to. Although the calculus proved very practical, purists were rightly concerned with the apparent division by zero in derivatives even though the indicated division was not by zero per se, but to a number arbitrarily close to zero.

You may know that division by zero is an absolute no-no in mathematics—inadvertently doing so with an expression which would evaluate to zero can lead to *proving* absurdities. If you divide one by one-tenth the answer is ten. And if you divide by one gazillionth, the answer is one gazillion. So the closer you get to dividing by zero the larger the number that results. If you actually divide by zero, the answer would be infinity and infinity is not a number that you can deal with using the usual arithmetic operations. A child knows that forever and a day or forever plus a day means no more than forever. This all led to the mathematical area named analysis that provides precise definitions of limits.

Domain is the term referring to from where an independent variable is taken, say the set of real numbers and *range* (or *co-domain*) to where the associated dependent variables resides, again say the set of real numbers. Either may involve sets of complex numbers or geometrically planes, solids or numbers depicting any number of dimensions—physical, normally envisioned as 3 or 4 but mathematical, no limit.

Collections of independent variables rather than just isolated variables can be mapped from any domain to any range in which case the mapping to the range may be called a figure. One of the simplest figures on a plane is a circle, the collection of points equidistant from a point called the center. Such a figure could be generalized to any dimension—say a sphere where the range is in three dimensions or to a hypersphere where the range is in any arbitrary number of dimensions.

A *partial derivative* of a function of several variables is its *derivative* with respect to one of those variables, *with the others held constant* (as opposed to the *total derivative*, in which all variables are allowed to vary). Partial derivatives are used in *vector calculus* and *differential geometry*.

A term endemic to integral calculus is, you guessed it, *integral* and its operation integration, indicated with an elongated s, $\int$. Integration is frequently thought of as the sum of the infinitesimal areas bounded by a bounded curve on a plane.

Calculus

Where algebra has practical applications outside of science and engineering—although the most frequently used mathematics there as well—at least an understanding of the concepts and acquaintance with the methods of basic calculus is essential for the understanding of the mathematical developments in the physical sciences and engineering.

The credit for the invention of the generalized differential and integral calculus is shared by Isaac Newton and Gottfried Wilhelm Leibniz although methods for specific solutions had been around for centuries [such as the method of exhaustion for determining the area bounded by a curve by finding the limit approached by a circumscribed polygon as the number of its sides indefinitely increase].

The realization that integration and differentiation are inverse operations is extremely interesting as well as extremely useful in practice. It is known as the *fundamental theorem of calculus*.

The *differential calculus* is a general methodology used to determine the instantaneous rate of change, the so-called derivative of a variable, at a given point.

Apart from the symbolic mathematics the differential calculus provides means to determine the instantaneous rate of change of a variable at a point, one way it can be determined is from the curve the variable describes when plotted in Cartesian coordinates. It is the slope [or tangent] of the curve at a point, that is, rise-over-run where rise is the value of the value of the vertical coordinate, traditionally y. And run is the value of the horizontal coordinate, traditionally x. This determination proves extremely useful in so many applications that I am departing from my overall avoidance of mathematical notation to illustrate its use here.

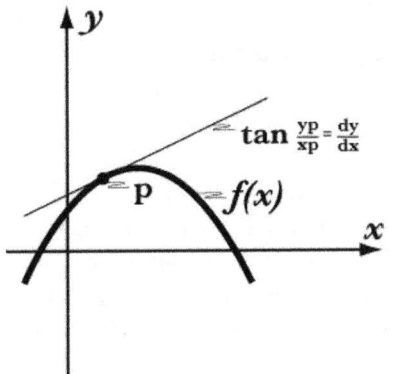

Figure 2-3 The derivative, dy/dx, at point P on $f(x)$
The curve here is represented as $f(x)$, the tangent to the curve at point P is denoted by $\tan f(x)$ [at P] $= y_p/x_p = df(x)/dx$ [at P]. That quantity represents the derivative of the curve with respect to x at the point P as calculated in the manner described in following text. Because $f(x)$ at P equals the value of y at P here, $df(x)/dt$ might be spoken of as dy/dt and be *okay* here although such is not the best practice.

You may choose to visit giac2002.wordpress.com and under the BOOK LINKS heading, put your mouse pointer on the word "here" associated with link 3, hold down the CTL key and click to view an outstanding dynamical presentation exhibiting the geometric meaning of the derivative as one changes the point of tangency to a curve.

You may also choose to visit giac2002.wordpress.com and under the BOOK LINKS heading, put your mouse pointer on the word "here" associated with link 4, hold down the CTRL key and click to read an outstanding detailed pimer on derivatives including graphics.

If the curve is defined as a function of x, "$f(x)$" and Δx is defined as a change in the x dimension direction, its instantaneous rate of change is the limit of $f(x + \Delta x) - f(x) / \Delta x$ as the limit of Δx approaches zero. This limit is conventionally denoted as $df(x)/dx$.

Examples:
If $f(x) = x^2$, then $(x + \Delta x)^2 - x^2 = x^2 + 2x\Delta x + \Delta x^2 - x^2 = 2x\Delta x + \Delta x^2$ that divided by Δx equals $2x + \Delta x$ that when Δx approaches zero as a limit, approaches $2x$, then $2x$ is called the derivative of x^2 and the derivative of $f(x)$ when $x = 2$ would be 4.
By a similar line of reasoning:
If $f(x) = x^n$, then the derivative of $f(x)$ would be nx^{n-1} where n is any integer.
If $f(x) = x$, the derivative of $f(x)$ would be zero for any x. That is, the slope of any constant x is zero.

Derivatives, also frequently referred to as *differentials*, [somewhat more accurately phrased, a derivative amounts to the ratio of the differential of a dependent variable divided by the differential of its associated independent variable] of any other function can be determined in a similar manner. A practitioner can determine a derivative on-the-spot, rely on his or her memory of what that derivative is or, failing either, look it up in available tables prepared for that purpose.

It is apparent that where any arbitrary curve has a derivative equal to zero at a point, that point is either at the greatest value, *maxima*, or the least value, *minima* of that curve. The derivative of any constant is zero.

The *integral calculus* is a methodology that can be used, among many other things, to determine the signed area between a curve describing the values of the independent variables of a function and the horizontal coordinate axis within the limits covered by the function's dependent variable. If the total area circumscribed includes areas below the coordinate axis, those areas are subtracted. This usage is restricted to what are called definite integrals.

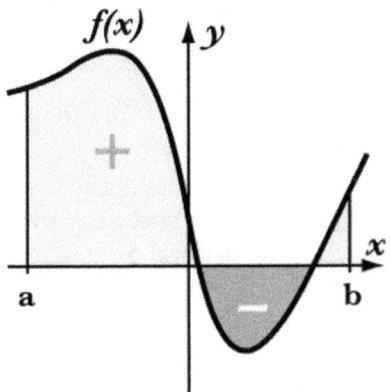

Figure 2-4 Representation for the definite integral of the function f of x between the limits a and b.

The curve here is represented as f(x). The value of the definite integral in this case would equal the area bounded by the curve portion below the horizontal axis subtracted from the area of the curve portions above the horizontal axis and between the vertical lines a and b. In this case this could be referred to as "the definite integral of the function f of x between the limits a and b."

You may choose to visit giac2002.wordpress.com and under the BOOK LINKS heading, put your mouse pointer on the word "here" associated with link 5, hold down the CTRL key and click to view an excellent more detailed presentation of features of the integral calculus including very helpful colored graphics.]

In the more general case, that with indefinite integrals that include an indefinite constant displacement above or below the horizontal coordinate axis, the just described area determination would obviously only work without modification if the assigned constant were set to zero.

Determining an integral by invoking the inverse of differentiation alone can only show the shape of a curve determined from the tangents to the points determined by differentiation. So alone, it requires the addition of a constant to describe absolute as opposed to relative values of the integral curve. Other information in a given situation is needed to meaningfully assign the constant, or offset to the abscissa of the integrated curve.

Real Mathematical Analysis

Later mathematicians were disturbed by what they considered the loosely described concepts of limits used in the development of the calculus and the concept of continuity of a function, sometimes described to students as a function with a curve that could be drawn without lifting a pencil from the page. The subject precisely defines other concepts and further precisely describes symbolic calculations with real numbers.

It defined a limit of a function at a point as one in which for any named change or less in the independent variable that was greater than zero, regardless of how small, a value greater than zero for the dependent variable could be chosen for which the change in the function would always be less. It defined continuity of a function at a point to exist if the value of the function at the limit point equaled the above described limit.

Expressed more formally (with the standard symbol < for less than, > for more than and quantifiers used in the predicate calculus [*epsilonics*]) a function of x, ($f(x)$ defined on an interval I is said to approach a limit A as x approaches a where a is a point of I, if given any or for every or for all (customarily designated by an upside-down capital A, $\forall$) (epsilon (ε) > 0 there exists ($\exists$) a delta (δ) greater > 0 such that ($\ni$) $|f(x) - A| < \varepsilon$ whenever $0 < |x - a| < (\delta)$.

Expressed more formally (with the standard symbol < for less than, > for more than and quantifiers used in the predicate calculus [*epsilonics*]) a function of x, ($f(x)$ defined on an interval I is said to approach a limit L as x approaches b where b is a point of I, if given any or for every or for all (customarily designated by an upside-down capital A, $\forall$) (epsilon (ε) > 0 there exists ($\exists$) a delta (δ) greater > 0 such that ($\ni$) $|f(x) - L| < \varepsilon$ whenever $0 < |x - b| < (\delta)$.

Vectors

Mathematical vectors are directed quantities which have proven very useful for describing and predicting results in physics. They may be represented as an arrow with a given magnitude and direction or by their component projections on coordinate axes.

You may choose to visit giac2002.wordpress.com and under the BOOK LINKS heading, put your mouse pointer on the word "here" associated with link 6, hold down the CTRL key and click to view an alternate vector discussion.

As complex numbers can be expressed as numbers as well as vectors, they have been previously mentioned here in the first mathematics section on counting. The sum of vectors can be described by the sum of the magnitudes of their corresponding projections on each of the coordinate axes.

Vectors may be simply position vectors that can be described in rectangular coordinates as so many units to the left or right, up or down and in or out. Other vectors include velocity, acceleration and force vectors.

Vector multiplication can be what is called a dot product, an inner product or a scalar product or what is called a vector product, an outer product or a cross product.

You can determine the value of the so-called dot product of two vectors by multiplying the value of the projections of each vector on the same coordinate axis and adding them. This product is a scalar. Another method is to multiply the magnitudes of the two original vectors times the cosine of the angle between them (determined from trigonometric tables).

You can determine the cross product of two vectors by multiplying the value of the projection of one vector on a coordinate axis times that of another vector on a different coordinate axis times the cross product of the unit vectors corresponding to each coordinate axes. The unit vector on the traditional positive x, y and z axis are i, j and k along the positive x, y, and z axes. Their cross products are defined as $i \times j = k$, $j \times k = i$, $k \times i = j$ or if the cross products are taken in the opposite orders $j \times i = -k$, $k \times j = -i$, and $i \times k = -j$. The cross product of vectors in the same direction is zero. Alternately, the cross product of any two vectors is the product of their magnitudes times the sine of the angle between them. In all cases, positive product vectors are at right angles to the two vector factors in the direction that a screw would travel if the first vector factor were turned clockwise towards the second factor.

A vector can also be described as a matrix with a single row or column or a first order tensor.

Imaginary Numbers, Complex Numbers, the Complex Plane and Hyper Complex Numbers

An *imaginary number* is nothing more than a so-called real number multiplied by the square root of negative one [that has no *real* solution].

The real part of a complex number may be represented geometrically as a point on a plane on the right from a point called zero, or the origin, and denoted by a positive sign or by a point represented geometrically as a point on a plane on the left of a point called zero, or the origin, and denoted by a negative sign.

The imaginary part of a complex number may be represented geometrically as a point on a plane upward from a point called zero, or the origin, and denoted by a positive sign or by a point represented geometrically as a point on a plane downward from a point called zero, or the origin, and denoted by a negative sign. You may choose to refer to figure 2-4 again.

In such a geometrical presentation, a complex number is either simply an imaginary number, a real [a usual, not imaginary] number or the vector sum of an imaginary number and a real number. The imaginary part is frequently represented by preceding it with the letter i.

The complex number plane or so-called complex plane is a plane consisting of all possible complex numbers so indicated.

If a so-called imaginary number unit vector were just called *up* or *down* as it is typically represented on the so-called complex plane, imaginary numbers would not have acquired the mystic connotations that can lead to unnecessary understanding difficulties.

An arbitrarily large number of other unit vectors, including among them ones representing *front* or *back* that might be represented by k and $-k$ can be postulated perpendicular to that unit vector in a hyper complex numbering system although they have only mathematical but not observable physical existence in our three dimensional space. The sum of any such number is representable by the sum of its components.

Quaternions

The mathematician and physicist William Rowan Hamilton came up with numbers he called quaternions in 1843. They consisted of real numbers and real numbers multiplied by the symbols i, j. or k.

In that system, the real numbers are not customarily preceded by any symbol. Those numbers preceded by the symbols i, j and k can be conceived as representing three perpendicular directions to the real numbers and if further preceded by a negative sign in the opposite directions. The product of any of these times itself, as in the complex number system for i alone, is a negative one. This overall system is thus four dimensional.

Alternately, one could conceive the real numbers being perpendicular to all of the other three dimensions [whatever that might mean]. i here could be represented as a unit vector up. The unit vector j could be represented here as a unit vector to the right and the unit vector k as a vector to the back. If preceded by a negative sign, the directions would be down, left and back.

Quaternions abide by all the operational rules of the algebra of real numbers except for order and multiplicative commutativity [that is, the product of two quaternions is not independent of the order in which they are taken—taken in one order, the magnitude of their product is the same as in the other order but it has the opposite sign]. The rules for multiplication of quaternions include those described for cross products of vectors. As always, going to Google.com and entering a term that you may not be familiar with can be worthwhile.

Hamilton spent the remaining 22 years of his life, predominantly, finding ways to apply quaternions to physics. But they proved so cumbersome that quaternions as a whole became abandoned until the latter part of the 20th century. Nevertheless, earlier their dot and vector product processes were adapted and used extensively and fruitfully, including for calculations in electromagnetism.

Quaternions represent one instance showing that however beautiful and elegant a mathematical method may be, it is not necessarily useful or even representative of the real world in practice.

Tensor Analysis

Complex or hyper-complex numbers such as quaternions can be represented as tensors. The vector concept can be generalized as tensors of various ranks [or orders or degrees]. The lowest rank, zero, is a scalar, then a tensor of the first rank is a vector.

An example of a second order tensor is its use in defining stress. That tensor is the (dot) product of the magnitude of the force acting on a vector representing the normal to an area with a magnitude equal to the area.

As noted in the "Understanding the Foundations" section, the methodology of tensor analysis provided the means to easily manipulate the 256 equations with 256 potential variables in each that were involved in deriving the curvature of space due to gravitational fields that Einstein postulated for his theory of general relativity.

The mathematical use of tensor analysis in the development of general relativity is excellently described for a layman in the reference, *The Einstein Theory of Relativity* by Lillian R. Leber.

The basic notation of tensor analysis is a wonderful exposition of the role of mathematics in providing succinctness. At its heart is the generalization of variables such as y, z and u with subscripts such as x_1, x_2 and x_3. This simple idea allows operations to be generalized by assigning a variable to subscripts and naming the range of numbers over which that variable is to apply.

The basic tensor notation involves the above as well as the standard mathematical designations for partial and total differentiation and summation to describe generalized transformations between coordinate systems. Einstein noted relationships that removed the need to explicitly show summation signs so those signs are no longer included. Succinctness is desirable because the human mind is limited in the number of elements it can deal with at one time.

As in any field of mathematics, the field of tensor analysis delves into the relationships that can be found among tensors thereby providing special tools for their manipulation. This, at the time less than two-decade-old field, was instrumental in changing Einstein's mind to one of great respect for the role of mathematics in physics as opposed to physical reasoning. He, with the aid of Marcel Grossmann, ended up being largely guided by tensor analysis in his development of the general theory of relativity,

Probability

The mathematical development of probability centers on the idea that a numerical probability is the number of so-called successes divided by the total number of attempts. Obviously, many more generalizations are derived in the general theory.

Differential Equations and Transform Methods

Unlike use of a straight-out formal method of *turning a crank* or referral to tables to spit out solutions to equations involving differentials as those that are available in many of the more elemental mathematics, different varieties of other than the simplest differential equations require different tricks to find solutions. Fortunately, there are transform methods that can often do the trick relatively simply—Fourier and Laplace transforms to name two. In these cases the original equation is converted to a transform which is solved by transform methods and then the inverse transform is taken to arrive at the solution of the original differential equation.

A student proves many of the most useful forms of transforms and inverse transforms, memorizes them for later usage or in later practice uses tables of them that he or others have produced.

Differential equations are of two general types—ordinary and partial. Partial differential equations involve partial derivatives—that is derivatives with respect to each of two or more variables.

Other Geometries

Euclid's plane geometry as its name indicates is applicable only to figures on a plane. In the last few centuries other totally mathematically consistent geometries have been developed for which, understandably, many of the theorems of plane geometry do not apply. Two such geometries are spherical and hyperbolic. Both of these geometries drop Euclid's parallel line postulate.

In spherical geometry, the shortest distance between two points is not a straight line but a great circle and great circles do intersect. Spherical geometry is directly applicable to the earth insofar as the earth is a mathematical sphere and so is very useful. On the earth, plane geometry is only approximately applicable over limited areas.

Boolean Algebra, Digital Logic, Digital Arithmetic and Digital Computers

Boolean algebra was invented by George Boole. It assigns a 1 for a given or true statement and a 0 for a false statement in a digital logic that requires only a limited number of logic relationships.

In general purpose electronic digital computers, ones or zeros are differentiated by two different defined voltage levels that could be set with such as a physical switch, but most likely by electronic gates.

The computer sequence to add two numbers A and B might be to get the value of a binary number from a user input and store it in memory location A. Then put that in association with the value previously stored in memory location B. Then put these as inputs for the addition operation procedure previously stored in the computer memory location C. Finally, execute the addition operation procedure and put the result in memory location D for a potential output for printing or another output operation.

The basic logic relationships include *and, not, or* and exclusive *or*. Corresponding electronic logic elements have been built for use in so-called logic design—including *and, or, not, nand* and *nor* gates that can be wired together to produce an output voltage corresponding to the truth or falsity of any logic statement.

Nand gates for instance yield an output voltage indicating a one if two input voltage levels indicating a one are NOT set; *nor* gates yield an output voltage indicating a one if NEITHER of if two input voltage levels indicating a one are set. Obviously hooking the output of a *nand* gate to a *not* gate will yield an output level from that gate indicating the same as a single *and* gate—that the inputs to both of the input levels to the *and* gate were set to a voltage level indicating a one. It is sometimes more efficient to use *nand* or *nor* gates than the more obvious *and* or *or* gates.

Such gates are used for special logic design or in the design of the logic functions of general purpose digital computers. The basic if-then function of these elements is subject to the same philosophical limitations of any crisp logic, due to its principle of the excluded middle. That is, given an input that is declared true any output of an *if-then* statement will be declared true. For instance, given the *if* input that the moon is made of green cheese then any if-then statement concerning Superman's birthplace will be declared true even if it is literally false. This is one origin of the computer usage truism—*garbage in, garbage out.*

The arithmetic functions of general purpose digital computers use binary numbers.

All numbers can be represented with respect to any base. Our every-day numbers use the base 10, Babylonians used the base sixty, binary numbers use the base 2. Most simply in binary arithmetic, the sum of 1 plus one equals 10—that is a unit in the *two's* place and no unit in the *one's* place. Counting in a binary sequence equivalent to zero to eight in decimal is 0, 1, 10, 11, 100, 101, 110, 111 [1 in the 4's place, one in the 2's place, 1 in the 1's place sums to 7], 1000.

A program is input into the computer as to how to proceed for a particular operation. The computer's basic sequence once initiated is to either access new data or access the data in the memory location given by the program to determine the previously set data at that memory location and to access the computer memory for the elemental operation to perform and where in memory to store the result.

A description of basic input, output and memory general purpose digital computer devices and their evolution an entirely additional different matter, not, at this writing at least a subject for this book. They have included manual switches, punched cards, magnetic and optical tape/disks.

A basic computer operation can be completed in the order of millionths of a second or less so computers can perform operations very much more quickly than a human. It is generally more efficient to calculate the values in tables, for instance, such as those of trigonometric functions rather than retain the tables in computer memory. Also including extensive error checking that can be done at *light speed* puts computer output accuracies far beyond that typical for a human.

Human involvement remains necessary to write computer programs as it is necessary to set up mathematical equations to be solved. But computers will solve the corresponding programs with much less chance of error than a human *cranking out* a mathematical equation.

Human memory was substantially re-enforced with the printing press and books. But once information is stored in computer memory, it can be much more rapidly accessed.

The knowledge that was once stored in human brains and books is now largely stored in computer networks and can be readily accessed by what has become known as googling.

Humans can still understand each other better and fill in the gaps that a question may leave open without the questioner having to guess the required effective literal follow-up questions. And humans can intuit the desirable follow-up operations when there is an unexpected computer output not covered by the original programming.

It is impossible as a matter of practice fo prior programming to cover the universe of all possible human foibles or unexpected computer outputs.

Chapter 3
Electricity and Magnetism with Hydraulic Analogies

Introduction

All electrical realities follow from the fact that like charges repel and unlike charges attract each other in direct proportion to their charge and the inverse square of the distance between them—and that magnetic effects result from charge motion, due to special relativity.

There is no denying the practical efficiency in engineering development of using mathematical equations that have been proven experimentally. Those convinced of the accuracy of mathematical procedures can be convinced of the reasonableness for further detailed physical results that they prophesy especially when experts testify that the results agree with observations. In this fashion, I was convinced of the truth of the basic equations of special relativity when I saw that they flowed seamlessly from basic very reasonable postulates and straightforward high school algebra.

Analogy, on the other hand, has been given a bad name in some quarters, because it can lead astray. However analogies frequently lead to understanding and insight. It is foolish to ignore previous knowledge. Analogies between known or accepted facts or theories can lead to at least preliminary acceptance of newer beliefs. They may lead to speculations which in turn may lead to valuable hypotheses for further
fruitful investigation.

I believe this all is especially true with hydraulic analogies to electricity. They can be useful in arriving at an intuitive understanding of the physical principles and interrelationships involved and the understanding of such key concepts as resistance, capacitance, inductance, and electromagnetic waves.

Physical Fields

I'm tempted here to introduce a novel term: *stortions*, to specifically refer to physical fields. Such fields are considered to be conditions of empty space induced by one of the four known fundamental forces—electromagnetic, gravitational and weak or strong nuclear forces. Science-fiction calls them space warps or they could be called distortions of empty space although both terms might imply these completely natural conditions as unnatural.

The magnitude of any of these fields at a particular point in space and time is determined by the amount of reaction of a test particle of a kind that reacts to the particular field type. The test particle must be chosen to be sufficiently small so that it does not change the nature of the field. For instance, the so-called force of gravity upon a particle depends on both the mass of the source and the mass of the particle, so to properly measure the field you need a sufficiently small test particle that its effect on the result is negligible.

It can be disturbing to consider that different fields affect only specific classes of objects. That is electromagnetic fields have no effect whatsoever on non-charged matter, nor do magnetic fields interact with stationary charged particles in the same inertial frame. This just brings home the fact that we can only describe how things interact. We currently can not explain the causes of gravitational or electromagnetic effects other than what causes the effects and what the effects are.

The foremost pioneer electromagnetic scientist, Michael Faraday, first introduced the concept of such fields as physical realities because they obviously carried energy. James Clerk Maxwell used the concept in his mathematical theory resulting in his unifying electromagnetic theory that deduced that electromagnetic waves traveled at the speed of light.

Maxwell, as many scientists, originally believed the waves were carried by a *luminiferous aether* because they resisted the idea that they could be a condition of empty space. Other known physical wave effects required a medium. For instance, even though you could not see a leaf move without visible contact with matter, it was known that the cause was the motion of air molecules in contact with the leaves. But it became established that no physical medium could be detected to support electromagnetic waves.

Now the concept of *real* physical fields is so well established that they have put the phrase *action at a distance* in disfavor in physics' circles even though in every-day language *action at a distance* is exactly what we are talking about when we are talking about physical fields. The distinction is that a physical field is what is considered in contact with the distant object that a field affects.

Einstein assumed that gravity consisted of physical fields in space in his general theory of relativity. The standard model of quantum theory considers fields to be made up of distinct corpuscles or particles that each have wave-like attributes. Gravity quanta are called gravitons, electromagnetic quanta are called photons. You can google information about these if you wish.

An excellent popular scientific book describing such is one by Richard Feynman, *QED* [Quantum ElectroDynamics], Princeton, New Jersey, Princeton University Press, 1985. One last diversion: some theorize that *empty* space consists of virtual particles that flash in and out of existence so quickly that they can not be detected. You could conceive of these as particles of the discarded aether concept. [Just thought I would throw this in for kicks.]

Electromagnetic waves are perceived to be the result of a kind of domino effect—the change of an electric field induces a slightly spatially displaced magnetic field at right angles to the electric field that is then followed in time by a spatially displaced magnetic field at right angles to that electric field, etc.

Maxwell's equations show that the so-called velocity of that effect is the same as that of light. One of the postulates of Einstein's theories is that no wave or other effect can travel faster than the speed of light. This postulate has never been shown wrong.

Once a field is established however, it may remain indefinitely. This last will turn out to be important in understanding the Ghosh unified theory of gravity and inertia. The inertial field resulting from the composite mass of the universe is overwhelmingly made up of the mass of the far distant stars millions of millennia in the past and so if will remain indefinitely. In this fashion, the Ghosh theory inertio-gravitational field behaves just like the absolute space that Newton had to reluctantly postulate for objects to move relative to in order to explain inertial effects.

Hydraulic Analogies

Hydraulic analogies to static Coulomb force and moving Coulomb force interactions between electrons in a stream make most electrical effects understandable. The major exception appears to be magnetic effects and hence any inductance effect which is commonly attributed to magnetic forces. As it turns out inductance effects are analogies to mechanical inertia effects. It would be possible to attribute inductance to simply the inertia of electrons in a stream except for the lack of significant electron mass and the almost imperceptible gravito/inertial interactions at molecular levels.

It should be understood that hydraulic analogies are just that--analogies. Some hydraulic analogies used in this book use fluids as analogies to electric charges and electric charge flow inside flexible external hoses that are analogous to electrical conductors. Some fluid analogy simply represents empty space, [that may be referred to as an ocean], and its contained electromagnetic fields. The use of the fluid ocean analogy is meant to help you gain an intuitive understanding of electric fields and electrical wave propagation across space that has been altered by electric charges and charge flow.

The use of a fluid ocean analogy for space can be a great help in understanding and deeper comprehension of the interactions between electrical charges; however, like all analogies, it does not capture everything. Like Zen masters say: "the finger pointing at the moon is not the moon." The fluid ocean analogy is similar to what used to be called the aether, but it was found impossible to detect an aether separate from empty space. In that sense, the aether turned out to be simply an analogy for what observations showed actually happened.

Einstein tried but was unable to provide a mathematical basis for treating electrical interactions like he was able to do for gravitation-mass interactions—that is, changes of what is called the *time-space continuum*. However interactions of charges upon other charges or moving charges upon other moving charges demonstrably occur, and can be precisely predicted using established mathematical models.

Hydraulic Analogy to Electron Flow and Resistance

Fluid flow in a hose with a flexible exterior within a fluid is closely analogous to electron flow in a conductor because the outer shells of the atoms making up a fluid's molecules are electrons and to that extent interact with each other similarly as if the molecules were electrons. The basic differences in this regard are mass, size and the apparent lack of an attractive magnetic effect in hydraulics. The effect of electron mass is infinitesimal compared to the effects of its charge and atoms and molecules are considerably larger than electrons. Overall the analogy of simple hydraulic considerations to electricity has an excellent physical foundation.

The opposite is definitely not true especially for those with an intimate knowledge of hydraulics. Real fluids do not have the idealized characteristics we assign in our analogy. For instances: hydraulic resistance due to friction on hose walls is nowhere near as simple as electrical *ohmic* resistance, *cavitation* has no electric analog and turbulence which has an analogy in electrical eddy currents is not considered in our straightforward electrical analogy.

The intuitive aspect of our analogy tends to be related to fluid flow similar to that from garden hoses.

You may choose to visit giac2002.wordpress.com and under the BOOK LINKS heading, put your mouse pointer on the word "here" associated with link 7 and link 8, and in each case, hold down the CTRL key and click to find more information on fluid flow theory. [To access and read these files, you need to have the free Acrobat Reader for PDF installed and operational on your computer.]

We have a direct familiarity with basic fluid flow. Early on people called electricity *juice*. Copper wires and pipes or hoses were seen to be analogous to wire conductors; they have to be connected for fluid or electricity to flow.

Where mass is a characteristic of all matter and energy, electric charge is also for anything not electrically neutral. We have come to accept that matter is characterized by atoms composed of electrical charges as well as mass. The nucleus of all atoms is positively charged and as we have found that opposite charges attract, the nucleus attracts much smaller freely-moving electrons which are negatively charged. The nucleus is positively charged because its make-up includes positively charged protons.

Not of particular interest here, but we do know that like charges repel so the protons in the nucleus would not stay together if it were not held together by a short-range more powerful force, called the strong force.

Under certain circumstances the so-called *free* electrons can escape the atoms' nuclei and be acted upon by external electric forces to flow in a conductor. It turns out that copper and gold are two of the materials with the highest conductivity—that is, these are the materials which have the least opposition to electron flow. Copper is most used because it is much less expensive.

Atoms as a whole, including those in fluids, resist occupying the same space at the same time because the electrons in their outer shells repel each other. So, if any atoms in a liquid are pushed they will push other atoms in the fluid. For the same reason, if any electrons in a stream are pushed they will push other electrons.

As you likely know, the term physicists, including applied physicists, engineers, use for push or pull is force. The force on a fluid across a section of hose, and so the amount of fluid flow, is proportional to the longitudinal pressure difference across that section of hose. Pressure is precisely defined as the force per lateral area of the fluid in the hose. Everything else being equal, the greater the pressure difference across a given length of a constant diameter smooth unobstructed hose, the greater the fluid flow momentum (mass times speed).

Similarly the force on an electron stream is proportional to the voltage difference across a length of conductive wire. Everything else being equal, the greater the voltage difference across a length of a constant diameter conductive wire of identical material, the greater the amount of electric current flow. This leads to the relationship of electrical resistance, current flow and voltage, the so-called *Ohm's law*. A given electrical resistance (R) [measured in ohms] with respect to the amount of electrical charge (Q) per second [measured in amperes (I)] is equal to the voltage difference (V) across the resistor, that is V = IR.

This relationship applies to any resistance to which it applies, that is *ohmic resistors*. Most materials of constant length and cross-sectional area have a constant resistance over a wide range of temperatures. So resistors can be manufactured and produced so they can be purchased to produce consistent calculated results.

Clean water hoses provide minimal resistance to water flow. However, regardless of how free from internal obstructions or roughness they may be, they provide some resistance. To make the analogy clearer, water-hose resistors with particular water hose lengths and diameters could be constructed to provide named resistances. Then such resistors could be put in line with each other to provide a total resistance equal to the calculated sum of the individual resistances. The same is true for electrical resistances. That is, where summation is indicated by the symbol $\sum$, total electrical resistance would be mathematically equal to $\sum R$ where R is the resistance of each individual resistor.

The pressure drop across a water resistor in series with other resistors in line with an overall water pressure source such as that provided from a water tower, would be equal to the proportion of that water resistor's resistance to the total overall water resistance of the series of water resistors. The same is true for an electrical resistor as to its voltage drop in series with other resistors wired to a voltage source such as a battery. That is the voltage drop (V) across such a resistor would be mathematically equal to R, t*he resistance of <u>each</u> individual resistor*, divided by R_T = the total resistance of the series of resistors times E = the voltage or so-called electromagnetic force (e.m.f.) provided by a voltage source across the series of resistors. That is, $V = E \times R/R_T$.

Obviously, if you put two sections of equally resistant water hose in parallel with each other so that both are subject to the same water pressure across them, twice the amount of water would flow through their combination. In this situation, it may be more obvious if you consider conductance rather than resistance where conductance is ease of water flowing through hoses [or electrical resistors]. The same would apply to electrical resistors in parallel, that is if they are equal, twice the amount of electrical current would flow through the combination than would through each individually if they had the same voltage across them both. And conductance would be a measure of the ease of current flow through wires.

Such a conductance measure is exactly the mathematical reciprocal of the measured resistance, that is $G = 1/R$, where G represents conductance and R is resistance. With these conventions, the total parallel conductance would be the sum of the individual conductances ($G_T = \sum G$ or $\sum 1/R$ where G and R are the individual conductances or resistances. This explains the mathematics. An electrical engineer would most likely just calculate the total resistance in terms of resistance values, that is, as the reciprocal of the individual resistances, that is, $R_T = 1/\sum 1/R$.

In such fashions, engineers can calculate the resistances of any combination of resistors in series or parallel. For more complicated arrangements, an electrical engineer would likely use Kirchhoff's obvious laws that apply equally to laminar (non-turbulent) hydraulic flow and electric current. In electricity they may be expressed as follows:

1. The total current in a closed circuits is equal in all parts of the circuit. That is, no energy is lost such as through radiation.
2. The total of voltages in a closed circuit is zero where the drops across resistances are considered opposite in sign to the voltage from sources such as batteries.

Similar ideas apply to impedances in alternating current circuits as separately considered.

The resistance of a resistor is dependent on the resistor material, directly proportional to its length and inversely proportional to its cross-sectional area. You may choose to visit giac2002.wordpress.com and under the BOOK LINKS heading, put your mouse pointer on the word "here" associated with link 9, hold down the CTRL key and click to access more information on resistors.

We can hardly leave this particular discussion without mentioning that electrical resistance, although constant over a range of temperatures is not so over all temperature ranges. An extreme example is when the temperature approaches absolute zero. There molecular motion due to heat approaches zero thereby lowering the resistance to electron motion so as to approach what is known as super-conductivity. Because of what is called *skin effect*, resistance increases at higher frequencies. You may choose to visit giac2002.wordpress.com and under the BOOK LINKS heading, put your mouse pointer on the word "here" associated with link 10, hold down the CTRL key and click for more information on skin effect.

From all this, you can grasp that the seemingly esoteric idea of action and reactions across empty space that, in the cases of charge and mass diminish rapidly—inversely as the square of the distance between objects—are really at the root of our ideas of things *touching* each other. For instance, the bottom of our shoes touching the ground really represents the balancing of the attracting gravitational force of the entire earth against the electrical repulsive force of the aggregate of the sub-microscopic atomic outer shell electrons in our shoes.

Hydraulic Analogy to Electrical Capacitance

A hydraulic analogy to electrical capacitance and an electrical capacitor could be a pliable obstruction across a water hose.

In so-called direct current, that is steady current flow, a capacitor can serve as a storage element. For instance, assume there is a hose at an angle to the one carrying the main flow of current connected to a capacitor with its other side connected to the ground. Then while pressure is applied so the water will flow it will also stretch the pliable obstruction, say a diaphragm. The amount of stretch if the diaphragm is not broken will serve as a memory of the amount of pressure that was applied.

When the pressure is released, the diaphragm will itself act as a source of pressure as it returns to its static condition. The amount of time that it takes for such a diaphragm to be distorted because of the water pressure or to return to its original state is determined by the value of the hydraulic capacitance (its capacity) and the hydraulic resistance in series with it to the original pressure source.

When current flow is not steady, such as is the case with back-and-forth alternating fluid flow and there is a hydraulic capacitor in line with that back and forth fluid motion, the stretching and unstretching of the diaphragm will pass the alternating fluid motion to the other side of the capacitor.

Capacitance

A diaphragm is an analogy to the dielectric in a capacitor, even a dielectric composed of nothing but *empty* space. An electric circuit symbol for an electric capacitor is two separated short parallel lines at right angles to the lines representing the wires to each side of the capacitor. The two short parallel lines are symbolic of the plates of a capacitor which might physically be parallel metallic surfaces separated by a very short distance by air or some non-conducting material called a dielectric—although more typically capacitor *plates* are surfaces wound around each other with a dielectric material between them and wires connected to the plates for connection to an external electric circuit.

When a steady voltage source initiates electron flow with a capacitor in series with the source in a circuit, electrons flow from the source over a wire connected to a capacitor plate where they build up until the capacitor is charged to the applied voltage or whatever voltage value is reached before the source is disconnected. Electrons on the plates on the other side of the dielectric are repelled during this process because like charges repel.

If a steady voltage source is disconnected from a capacitor, the capacitor will discharge with the contained electrons returning to the source and to whatever the wire connected to the opposite plate is connected. The voltage value over time when a capacitor is charging and discharging can be predicted in detail from the product of the value of a capacitor and the resistance in series with the capacitor. This has resulted in that particular product being called the RC time constant.

Any change in the voltage, including transients or pulses on one side of a capacitor will be passed to its other side and to the circuit elements, wire or ground to which it is connected. A capacitor is effectively a short circuit to rapid changes in a voltage applied to it while blocking any direct current fixed voltage. If the applied voltage to the capacitor is applied more smoothly up and down, the alternating current on one side of the capacitor will be effectively passed through whatever is connected to the other side although there will be some impedance to the alternating current flow through the capacitor.

Such impedance is called capacitive reactance. It's important to realize, the initial instantaneous impedance when an applied sinusoidal voltage is changing signs is very low and increases in one sign or another until the peak of the applied voltage of that sign is reached. Thus the sinusoidal current flow occasioned by the same source will lead that of a resistor by 90 degrees.

The capacitance value of capacitor is directly proportional to the area of its plates, their closeness and the properties of the dielectric material between the plates. Even free space has a dielectric constant of proportionality for whatever particular units of charge are chosen. It is called the permittivity of free space.

All electrical elements have some capacitive characteristic—between wires or other electronic components, wires and electronic components to each other or to ground.

You may choose to visit giac2002.wordpress.com and under the BOOK LINKS heading, put your mouse pointer on the word "here" associated with link 11, hold down the CTRL key and click for augmenting details on capacitance.

Magnetism

The foremost aim of this *Enhanced Magnetism Edition* as its title indicates is to enhance understanding of magnetism which the first edition freely admitted had no hydraulic analog that it otherwise heavily relied upon to provide an understanding of this and other electrical phenomena.

Magnetic interactions occur between moving electrons, magnets or by a magnet inducing magnetism in a magnetizable object made of such as ferromagnetic materials. This is more fully detailed later.

History

Magnetism was noted before around 600 BC as a property of an object, called a lodestone, or with *ferromagnetic* materials previously magnetized with a lodestone, that would attract iron. It was further noted around 1100 AD that, when such an object was allowed freedom of rotary motion, one end of such a material would point in a northward direction. Such north seeking ends became called the north poles of magnets and the other ends south poles.

It was further observed that the north pole of one magnet would repel the north pole of another magnet and attract the opposite end of another magnet, called the south pole—that is, opposite poles of magnets attract, while like poles of magnets repel.

Rare earth and Bitter magnets, discussed further below, represent some of the technological advances made in magnetism in the last century.

Magnetic Force

Magnetic force can be readily exhibited by running direct electric current through an insulated wire wound around an object made of ferromagnetic material. By doing this one would have constructed an electromagnet that can attract ferromagnetic materials or attract or repel a similarly contructed electromagnet—depending on the direction of its direct current flow and which of its ends are aligned with which end of the previously contructed electromagnet.

The strength of an attractive magnetic force can be most directly and obviously measured by the mass of a magnetizable object it will hold—the strength of weak magnets perhaps most simply by the number of interlocked paper clips they can lift. Of course, there are many commercially available measuring devices for magnetic effects.

The holding force of magnets depends on the magnet's material, shape and dimensions, the square of its flux density and any intentional or unavoidable gap between the magnet and the material attracted. Because of the latter, paint or grease on a magnet's surface can severely lower its holding force.

Ferromagnetic materials included iron, nickel, cobalt, carbon steel and other alloys before rare earth permanent magnets were developed in the 1970's and 1980's. Some of such permanent magnets are said to hold up to 1300 times their own weight. One rare earth 4" × 2" × 2" block permanent magnet is advertised as having a pulling power of 624.8 pounds.

Very powerful electromagnets with holding forces of tons can be built with large electric current conducting through thousands of turns of insulated wire. Ferromagnetic core only helps till it saturates—that is, until essentially all its magnetic domains [electron spins and orbits] are aligned. You may choose to visit giac2002.wordpress.com and under the BOOK LINKS heading, put your mouse pointer on the word "here" associated with link 14, hold down the CTRL key and click for an overall discussion of magnetic domains.

Increased currents produced by larger voltages across helix windings or superconducting conditions achievable at extremely low temperatures will further increase the magnetic strength. Even higher strengths are achievable with Bitter electromagnets invented in 1931 by Francis Bitter. You may choose to visit giac2002.wordpress.com and under the BOOK LINKS heading, put your mouse pointer on the word "here" associated with link 15, hold down the CTRL key and click for more on bitter magnets.

Electromagnets can be useful in transporting ferromagnetic or ferromagnetic encased materials and then releasing them without complex mechanical grasping and releasing mechanisms. People need to be very careful and not be under them because of the danger of their immediate release with a transient power loss.

Electromagnetism

Hans Christian Oersted noticed while preparing a demonstration of electricity for students during the winter of 1819-1820 that a steady electric current changed the direction pointed to by a compass needle.

He further found that, when a straight wire carried a steady current, magnetic field lines encircle the current-carrying wire.

In the wake of Oersted's discoveries, Andre Marie Ampere established within the year that that electrical current moving in the same direction in nearby wires caused the wires to attract if the currents were moving in the same direction and caused them to repel if the currents were moving in opposite directions.

Figure 3-1 Magnetic Lines of Force Encircling Electric Current Carrying Conductor

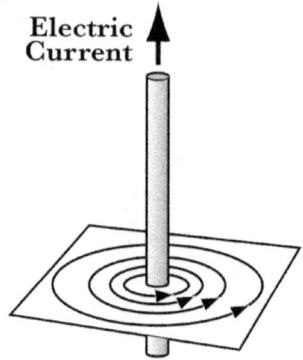

You may choose to visit giac2002.wordpress.com and under the BOOK LINKS heading, put your mouse pointer on the word "here" associated with link 16, hold down the CTRL key and click for further information on Ampere's experiments.

It stands to reason that the closer the wires the greater the magnetic effect and that the force between two current carrying parallel wires would be magnified if more wires were involved as when they are in a helical arrangement with coils in proximity to each other—in proportion to the number of nearby coil windings.

These inferences are indeed borne out by observations; the first can be referred to as an ampere turn effect where amperes are the measurement of the amount of electrical current flowing within the coils. All observations have established this.

The coils in which the current originates could be called primary coils. Those in which current is induced by a process that we will further detail could be called secondary coils.

In 1905 Albert Einstein deduced his theory of special relativity [ESR] which has survived all attempts to experimentally find any experimental objections.

Physicists have since established that magnetic effects are strictly due to electrical observations in different inertial frames. Edward Purcell specifically explained how ESR explains the magnetic effects between electric current carrying wires in his textbook: *Electricity and Magnetism* 2nd edition for a Berkeley physics course vol. 2.

According to Edward Purcell: in his book *Electricity and Magnetism* "the magnetic interactions of electric currents can be recognized as an inevitable corollary to Coulomb's law. If the postulates of [special] relativity are valid, if electric charge is invariant, if Coulomb's law holds, then, as we shall show, the effects we commonly call 'magnetic' are bound to occur."

To briefly summarize the hydraulic analogy of the first edition of our current book to what was really an analogy to moving Coulomb force* charge—it consisted of a flexible-exterior hose containing the moving charge whose ballooning action within a fluid and the fluids reaction to it was analogous to the field produced in space for which the analogy was the overall external fluid.

*The so-called Coulomb force is directly proportional to the product of two interacting electrical charges and inversely proportional to the square of the distance between them. Charles-Augustin de Coulomb published this *law* in 1785.

All Magnetic Force is Caused by the Motion of Electric Charges

The following six statements summarize this book's discussions of magnetic force:
1. The cause of magnetic force is always the motion of electric charges. It is said that a magnetic field is simply the electric field viewed from a different relative velocity as in Einstein's theory of special relativity.
2. Electrical currents in nearby wires cause the wires to attract if the currents are in the same direction or to repel if the currents are in opposite directions.
3. Einstein's theory of special relativity [ESR] explains how the motion of electric charges causes magnetic force.
4. A moving Coulomb force field induces other moving Coulomb force fields with the accompanying motion of their associated electric charges.
5. Moving charge fields are traditionally called *magnetic fields* with their handling mathematically firmly established although they could reasonably be called Moving Coulomb Force [MCF] fields.
6. Unpaired [and so not canceling] electron spins and orbits account for the miniature magnetic domains which make up the overall magnetism of magnetized objects.

Figure 3-2 Magnetic Field of a Bar Magnet and its Associated Spinning Electrons

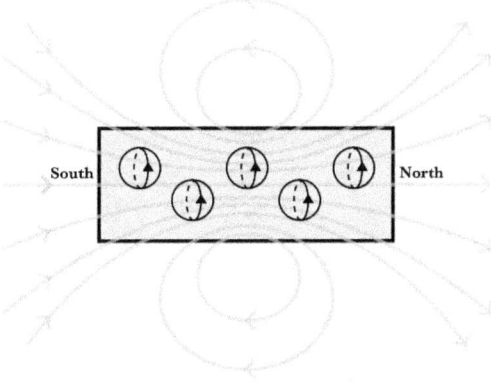

Figure 3-3 Bar Magnet Picking Up an Iron Nail

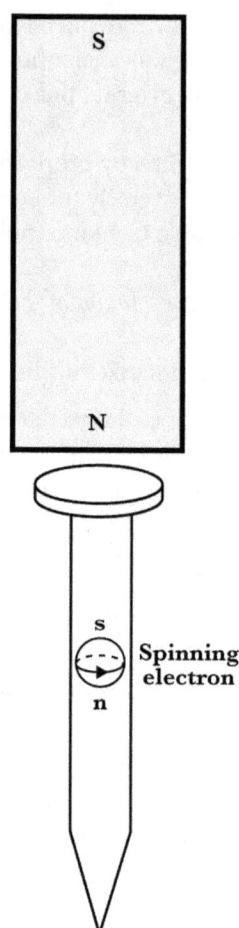

Statement 1. and 2. are simply reports of all reported observations.

Statement 3. is explained by physicists with Einstein's deduction, never found experimentally false, that the measurement of lengths diminishes as the relative speed of an observed object increases, and so the observed density of electric charge and its experimentally measured total charge increases with the increased relative speed of the charged objects. Nobel Prize winner, Edward Mills Purcell, explained this explicitly for the measured magnetic forces between nearby wires described in statement 2 as more fully detailed below under "Basic explanation for magnetism with Einstein Special Relativity."

Basic Explanation for Magnetism with Einstein's Special Relativity

The size of a proton has been said to be 10^{-15} meters. That of an electron is said to be less. That is to say there are 10^{15} that is 1 followed by 15 zeroes protons in a length of one meter (roughly one yard). In either case, the dimensions involved are extremely small so that many, many electrons or protons can be involved.

The key point is that the force between objects is proportional to the density of the protons or electrons involved. Einstein's theory of special relativity shows that the relative dimensions of any moving matter containing electrons or protons shrinks so that the charge density of the objects increases.

When electrons are flowing [or Coulomb forces are moving] in the same direction through two parallel wires, the protons in one wire are moving oppositely relative to the speed of the electrons in the other wire. Then because of the relativistic contraction of length, the effective density of the protons from the perspective of the electrons in the other wire increases and so their cumulative effective charge from the perspective of those electrons is positive. This then results in a magnetic attraction between the two wires because unlike charges attract.

On the other hand, if the electrons in the two wires are moving in opposite directions, their relativistic contraction in length results in a dynamically induced greater negative charge between the two wires. This then results in a magnetic repulsion between the wires because like charges repel.

Should this explanation not completely satisfy you, Ampere's original observations on the mutual magnetic attraction/repulsion effects between parallel wires have been so repeatedly verified without exception that they alone are a sufficient basis for a further comprehension of magnetism. The special relativity relationships provide an additional theoretical background of course.

The relations between parallel wires imply the same reaction whenever there is motion between mobile electrons and protons in any type of atom in matter. Thus starts an explanation for the magnetic force between electrons spins or orbits when bits of matter are very close or said to be *touching*.

This theoretical explanation has been supported without exception (similar to no counter-example or no contradictions in mathematics) for all physical observations—the ultimate designation of accepted *truth* in physics.

One might object that the very small average drift speed of electrons within a conductor would negate any such relativistic effects that are typically considered only significant at speeds approaching that of light. But Edward M. Purcell, Nobel Prize winner in magnetism correlates such drift speed with magnetic force effects and explains it by the enormous electron forces due to relative velocity in which there is no effective cancellation due to protons as there is in the rest case.

In Purcell's book *Electricity and Magnetism* Berkeley Physics Course Volume 2, Second Edition, he applies the well-known special relativity equations on measurement of length contraction by a relatively moving observer and derives the resulting force between electron charges moving relative to each other. It turns out to be proportional to the product of the current in the wire and to the velocity of a test charge.

This absolutely establishes that it makes no difference whether the current is composed of electrons moving at 99 percent of the speed of light or electrons in a wire executing nearly random motions with a slight drift. He further notes that using the established magnetic field equations completely and concisely describes the magnetic effects without the tedious and confusing bother of having to transfer back and forth between relativistic coordinate systems.

Electron motion in a conductor with no voltage applied is random, but applying a voltage causes a small average electron velocity or drift. In his book, Purcell gives an example of the force between two parallel electric current carrying copper wires 1 mm in diameter, 20 cm in length and 5 cm apart with a mean electron drift velocity of only 0.3 cm/sec using the standard number of conduction electrons per length of conductor and the standard electron charge. He shows that this calculates to an attractive force of 80 dynes.

He remarks: "Now 80 dynes is not an enormous force, but it is easily measurable... And here with v [velocity] less than the speed of a healthy ant, it is causing a quite respectable force! The explanation is the immense amount of negative charge the conduction electrons represent, charge which ordinarily is so precisely neutralized by positive charge that we hardly notice it."

Edward Purnell's explanation using Einstein's theory of special relativity, reduces the *mysterious* force of magnetism to the equally basically *mysterious* but <u>fundamental</u> electric force of nature that is a *given* based on its countless observations.

<u>Statement 4</u> was discussed in the first edition of the current book. Basically a hydraulic analogy for a moving Coulomb force field is a flexible-exterior hose filled with water in an ocean. If you apply a longitudinal pressure across a length of such a hose, the end on the side to which pressure is applied will first balloon and then the formed ballooning will move longitudinally away from the pressure source. The ocean water in the vicinity of the hose will follow along with the ballooning inside the hose and induce a similar motion in a neighboring flexible-exterior hose filled with liquid.

<u>Statement 5</u> is further discussed immediately below:

Is it Best to Name a Magnetic Field Separate from a Moving Coulomb Force Field?

An established mathematical field vector **B** is customarily [and perhaps confusingly] called *magnetic induction* [It appears confusing to me in light of what I consider a convincing hydraulic analogy to what I call an MCF (Moving Coulomb Force) field.] The complete **B** vector is expressed in 3 dimensions as either description of the fields should be [if they are indeed different fields] for their full predictive mathematical use—although only the component parallel to the direction of charge flow is needed to understand the basic principle behind their depiction.

Magnetic force is always caused by the motion of electric charge —that is, motion of Coulomb force fields. This suggests that a separately named *magnetic* field may be superfluous and might reasonably be dropped for simplicity in accordance with the Occam's razor principle—other than the convenience that its associated historical mathematical definitions provide.

Accordingly, I may refer to what have historically been called *magnetic fields* as MCF fields, [Moving Coulomb force fields] while still referring to their *effects* as *magnetic effects*. We could infer that moving Coulomb forces centered on electrons in a conductor could induce alignment of the magnetic domains of ferromagnetic materials such as that used in a compass needle. That is, using the customary analogy of electron orbits for energy levels, the moving Coulomb forces in a conductor could induce the changes in those energy levels that are said to, in addition to electron spin, account for magnetic domains. MCF fields in one conductor pushing electrons or MCF's in another conductor reasonably explain the induction.

The idea that an electron can be largely described as a Coulomb force field centered on the electron also suggests the particle wave duality that seems to describe reality.

To quote from the book *Physics* by K. R. Atkins: "The magnetic field **B** enables us to calculate that part of the force which depends on the velocity...(The quantity **B** is more properly termed *magnetic induction* and the expression *magnetic field* is traditionally used for a slightly different quantity. In a vacuum these two quantities are identical ...Looked at from this point of view, the magnetic field **B** is a mathematical trick which helps us to calculate the forces between moving charges. The observational fact is that moving charges exert forces on each other which depend on their velocities. With sufficient mathematical ingenuity we might explain this fact without ever introducing the concept of a magnetic field....One thing is certain: the concept certainly helps to simplify the discussion."

From our perspective in this book of understanding rather than predicting precise results, we can reasonably proceed with nothing more than noting that charges moving with respect to each other with some parallellism produce a force between them different from that due to the static Coulomb force and agree to call that force *magnetic*.

Parallelism includes two parallel circular loops of currents which are as much magnets as bar magnets are—they attract if the currents go around in the same direction, but repel one another if their currents go around in opposite directions. The magnetic force is directed towards the center of the loops.
Notably, we need only consider the non-perpendicular components of the velocities without concern of the overall **B** field definition for the purpose of understanding as opposed to precisely predicting what may be overall complicated field descriptions or effects.

The direction of magnetic fields because they are not important for understanding and techniques for determining the direction involving left or right hand rules or clocks are a needless confusion here as they are reversed if the direction of conventional electrical current or electron flow is used and are not discussed in this book. Concepts involving such as magnetic flux are detailed in the book: *Fundamentals of Electric Waves* by Skilling as listed with other references just before chapter 4.

Statement 6 is further discussed immediately below:

Electron Spins and Orbits May Be Positioned in an Object to Make Miniature Magnetic Domains and Produce a Magnetic Force

One is frequently cautioned that quantum physics does not support the idea of electrons circulating around atoms. Nevertheless, a model of circulating and spinning electrons provides an effective analogy for what has actually been observed.

Using that model, randomly oriented electron spins and orbits in a ferromagnetic object could be induced by external fields to be similarly oriented by proximity to an already magnetized object or by being within a helix circulating current around the object and so making it a magnetized object.

Magnetism Explained in a Nutshell

Summing it all up, Ampere's determination of the magnetic effects of parallel currents are explained by special relativity and are exemplified by the magnetic domains set up by electron spins and orbits parallelism.

Individual moving Coulomb field forces repel electrons in some atoms so as to move them from their initial states and cause a majority of them to synchronize as a group with the overall external interacting MCF field. The MCF field of that majority may form a similarly oriented field in a nearby object. That is, the nearby object may also become magnetized.

In any event, when any magnetized object is put in proximity to an object made of a ferromagnetic material, it will magnetically attract or repel the other object through the cumulative effect of its atomic electron spins and orbits in the same way that Ampere discovered that parallel electrical currents would attract if they were moving in the same direction or repel if they were moving in opposite directions.

Most materials show no net magnetic effect because for the orbit and spin of any of their electrons, there is another electron with the oppositely oriented orbit and spin so the magnetic effects cancel each other. Ferromagnetic and ferromagnetic materials are exceptions as they are materials in which there may be an excess of electrons spinning in the same direction on the same axis to produce a large combined magnetic field.

I have read that ferromagnetism arises primarily from the Pauli Exclusion Principle in conjunction with quantum mechanics spin and—that classical theories are unable to account for any form of magnetism. These topics are further discussed on the Web. Ferromagnetism and magnetic domains are specifically addressed in Web articles. You may choose to visit giac2002.wordpress.com and under the BOOK LINKS heading, put your mouse pointer on the word "here" associated with link 17, hold down the CTRL key and click for further information on ferromagnetism.

Magnetic Effect on a Magnetizable Object of a Current Carrying Helix without a Ferromagnetic Core

Up to this point, this discussion has centered on the situation where a physical magnet attracts or repels a magnetizable object or how inserting a ferromagnetic core in a current carrying helix forms an electromagnet. A current carrying helix without a ferromagnetic core can also act as a magnet. In fact when there are essentially no unaligned magnetic domains remaining available in such a core [that is, effectively no similarly aligned electron spins or orbits], the core is said to be in saturation. The only additional magnetic force that can be applied is that due to the current carrying helix alone.

Much magnetizing force may be applied without a ferromagnetic core in the current carrying helix as long as the material to be magnetized is magnetizable. This can be visualized as the moving Coulomb force fields in the helix alone pushing the electrons in the magnetizable object so as to further change their electron spins and orbits thereby increasing the total reacting magnetism of the magnetizable object and so the mutual magnetism.

This is a way that magnetism can be induced or much greater enhanced without having a magnetic core in the magnetizing current-carrying helix. As the holding strength of a magnetic field is directly proportional to the current in a helix, reducing the temperature of the helix to the superconducting realm [where thermal effects are so reduced that the wire resistance essentially disappears] increases the current and so the magnetic effect. Some of the magnets with the greatest holding strength are so-called *Bitter* electromagnets

Bitter electromagnets are constructed of circular electrical conducting plates separated by insulating spacers rather than coils of wire. Cooling water flows through holes in the plates. The electrical current flows in a helical pattern through the plates.

Francis Bitter invented these electromagnets in 1933. As of 31 March 2014, these were the strongest magnets on earth having achieved field strengths of 37.5 Tesla at room temperature. A hybrid device consisting of a Bitter magnet inside a superconducting magnet has since produced field strengths of 45 Tesla.

Nomenclature

Although I believe the above is adequate for a basic understanding of magnetism, there are other common terms used in dealing with magnetism that warrant discussion.

A definition of flux in my Webster dictionary is a flow of energy considered to be a fluid. This term could be used for any physical quantity typified as a fluid. A hydraulic analogy to flux is therefore a fluid and an analogy to the amount of flux is the quantity of a fluid.

Magnetic flux, commonly symbolized, Φ_m, is the *total* number of magnetic field lines passing through a closed surface. The existence of these field lines are commonly shown by scattering iron filings around a magnet or an insulated electrical-current carrying conductor. Two dimensional photographs commonly exhibit a two dimensional slice of these three dimensional effects. These lines of force are not actually physically countable although a number relating to the strength of the field is arbitrarily assigned; the term *magnetic flux* is simply a measure of their total effect.

Perhaps the most common reference to magnetic flux is in Faraday's law that relates accompanying momentary or varying induced voltages or electromagnetic forces to the rate of change of the magnetic flux passing through an area. This is often mathematically symbolized as $E = -d\Phi_m/dt$ where E is the electromotive force (EMF), commonly also called voltage.

The vector field, **B**, in addition to being called *magnetic induction* is also called magnetic flux density. The vector field, $\mathbf{B} = \mu\mathbf{H}$ where **H** is the symbol for another magnetic vector field that is called *magnetic intensity*. μ is the constant of proportionality for this instance—it is the degree of magnetism that obtains in response to a magnetic field. μ_0 typically symbolizes the permeability of free space.

The following Web URL tabulates other magnetic field terms and relationships: www.magneticshield.com/pdf/how_do_you_measure_magnetic_fields_in_gauss.pdf

Topics on magnetic fields and electric fields are more thoroughly discussed with helpful geometrical illustrations in the book *Fundamentals of Electric Waves* by Skilling as listed with other references just before chapter 4.

Hydraulic Analogy to Magnetic Force

A hydraulic analogy to magnetic force might be taken to be the increased gravitational attraction of masses and so streams of masses moving with respect to each other as deduced by Einstein in his theory of special relativity

Quantification and Practical Applications

I have repeatedly stressed in accordance with the title of this book that this book is meant to forward understanding of qualitative relationships. When it comes to applying them, quantitative relationships are essential, and they typically require the use of mathematics to predict precise results. Such equations and their relationships with other established mathematical relationships are used for engineering purposes similarly as electronic computers without the practical need for engineers to follow all their intricacies. Tables and transforms of certain mathematical relationships to others more easily handled are common.

Inductance

Inertia, Inductia and Inductance

Inductance is nothing but the inertia of electric charge, most basically the reactance of electrons opposing any change in their motion. In this respect, inductance might be more analogously named *inductia*.

The basic definition of electrical inductance is that its magnitude times the rate of change of the current flowing through it is proportional to the voltage across it. This is a strong analogy to the mechanical inertia effect which is the opposition to any change of mass motion. The usual extremely concise and precise mathematical relationships clearly show this.

In the case of mechanical inertia, the interaction is represented as the force, **F**, resisting the motion of matter as equal to the mass, m, times the change of its velocity, **v** [in speed or direction] $d\mathbf{v}/dt$ where t = time. That is, $\mathbf{F} = m \times d\mathbf{v}/dt$ [or, because $d\mathbf{v}/dt$ is defined as acceleration **a**, the most familiar $\mathbf{F} = m \times \mathbf{a}$].

In the case of inductance, the interaction is represented as the electromotive force, **V**, resisting the motion of charge, as equal to the charge, q, times the change of its velocity, **v** [in speed or direction] $d\mathbf{v}/dt$ where t = time. That is, the counter electromotive force [emf, measured in volts, $\mathbf{V} = q \times d\mathbf{v}/dt$.

This all might be summed up by observing that when an electric voltage is applied across a section of wire, the free electrons within the wire will move longitudinally and their accompanying electric field will move with them. When the voltage is removed the current flow would continue except that an electric field will be induced, as below, in the opposite direction that slows current flow.

Induction, self induction and inductance

The word *induction* as applied to electricity generally refer to three different instances:
1. Electromagnetic induction that refers to the transient electrical current effect induced by <u>alternating electrical currents or by insertion and removal of a bar magnet in a coil of conducting wire.</u>
2. Electrostatic induction that refers to the induction of electrical charge by the proximity of another electrical charge.
3. Magnetic induction that is caused by electrical current.

In the context of this book this would refer to moving Coulomb force induction. You may choose to visit giac2002.wordpress.com and under the BOOK LINKS heading, put your mouse pointer on the word "here" associated with link 12, hold down the CTRL key and click for further information on electromagnetic induction.

Self Induction or Inductance

Unlike mechanical inertia which is most frequently just an accepted property of inertial mass [though Mach and Ghosh lay it to the total mass of the universe], inductance as its name implies is generally accepted to be caused by *magnetic induction* rather than the moving Coulomb force induction suggested in this book's first edition. In my mind they represent the same phenomena, only under different names.

The induced voltage or electromotive force [e.m.f.] evident in *inductance* is negative feedback, that is an inverse e.m.f. to cause the opposition to changes in charge velocity. This is called Lenz's law in electromagnetism. As a generality, the opposite, to negative feedback is positive feedback. Uncontrolled positive feedback could cause destruction if not simply oscillation.

Mutual induction results when an alternating electron current flow in one circuit induces current flow in another circuit. Self induction is a similar phenomenon if one were to consider such as current flow in some links of an inductor inducing current flow in different links.

The self-inductive electric effect may be augmented by coiling a conductor carrying electric current around a ferromagnetic core. Commercial electric inductors or *chokes* are so constructed.

The inductance of a helical inductor is, in general, directly proportional to the number of helical turns squared, its cross-sectional area and the material of the core around which the inductor is wound. If the core is ferromagnetic, the inductance of a particular coil may be raised thousands of times.

You may choose to visit giac2002.wordpress.com and under the BOOK LINKS heading, put your mouse pointer on the word "here" associated with link 13, hold down the CTRL key and click for information about inductor construction.

The more complete term for electromagnetic inductance in a completely electromagnetically separate circuit is electromagnetic *self inductance* in distinction to the mutual inductance between electric circuits that would be separate except for magnetic coupling, e.g. electrical transformers.

A reader needs to be careful as to what a person or writer means when he or she uses the terms induction because either magnetic or electrostatic [Coulomb force] induction might be meant. [Possibly complicating matters even further, the magnetic field vector **B**, is often called magnetic induction as opposed to the magnetic field vector of magnetic intensity, **H**. [Except for ferromagnetic materials, $\mathbf{B} = \mu \mathbf{H}$ where μ is a constant called the magnetic inductive capacity of the medium].

It might be especially noted that the *Maxwell-Faraday* formulation expressed in vector notation **del** $\times$ **E** = the negative partial derivative of the instantaneous change of the vector **B** with respect to time is one of the four Maxwell equations that are fundamental for the classical theory of electro-magnetism. Vector notation is described under the heading "Vectors" in Chapter 2, Mathematics.

Inductance as a Circuit Element

In partnership with conductors, inductors have a time constant analogous to the RC time constant, the L/R time constant, except that it lengthens with a decrease in resistance, that is increase in conductance. That may be more readily understood with the hydraulic inductance analog. In that analog, the greater the current flow the greater the mass density flowing past a point so, by inertial effects, the longer it would take for an inductive effect to subside.

Inductors may be used as basic circuit elements to form resonant circuits with capacitors and for their separate properties in affecting inductive effects.

Resonant circuits may be formed by inductors and capacitors in series so that the inductive reactance and the capacitive reactance cancel so the only the resistance in a circuit remains in effect. This occurs because, unlike a pure resistance in which the voltage and the current are completely in phase, when the voltage across an inductor is highest, the alternating current through it is least while when the voltage across a capacitor is least the alternating current through a capacitor is the most. That is the voltages and currents are completely out of phase with each other: In an inductor the voltage leads the current while in a capacitor, the voltage lags the current. And at some, so-called, *resonant* frequency the capacitive and inductive reactance will cancel. An engineer or circuit designer is able to calculate a circuit's resonant frequency using the following equations for capacitive and inductive reactance. The capacitive reactance, Xc, of a capacitor of value C is one divided by the expression 2 pi times the frequency, f, times the capacitance, that is, $Xc = 1/2\pi fC$. The inductive reactance X_L of an inductor of value L is 2 pi times the frequency, f times, the inductance, that is, $X_L = 2\pi fL$. The total impedance except for the resistance in a series circuit is the square root of the quantity Xc minus X_L squared in that circuit. So the resonant frequency is that frequency at which $\sqrt{(Xc - X_L)^2}$ = zero, or more simply, when Xc - X_L = zero.

Hydraulic Analogies for Electricity and Inductance

The hydraulic analogies of electrical Coulomb force interactions are straightforward in many areas including electrical current flow, resistance and capacitance. But their description for the remaining circuit element and passive electrical characteristic, inductance, proves very involved. It turns out that magnetic effects, rather than Coulomb force effects are *customarily* those used in describing how inductance comes about. Even though it calls upon, perhaps one too many, causes--that is, magnetism which itself is nothing but the result of the movement of electrical charge--most basically electron, motion. I may hereafter choose to refer to the field as an *MCF [Moving Coulomb Force]* field and stick to only calling the effects *magnetic*.

Consider a flexible-exterior hose filled with water in an ocean. If you apply a longitudinal pressure across a length of such a hose, the end on the side to which pressure is applied will first balloon and then the formed balloon will move longitudinally away from the pressure source. The ocean water in the immediate vicinity of the hose will follow along with the ballooning inside the hose, but it will resist the ballooning motion and so resist the water flow inside the hose. That opposing reactance is a hydraulic analog of self inductance.

Self induction effects are magnified between coils in a coiled flexible-exterior hose immersed in a fluid. The pulsating balloons set up by some coils will be passed on to other coils through the fluid so inducing synchronous pulsating balloons moving in the same direction in the other coils although the induced balloons will be 180 degrees out phase with the balloons in the originating coils—that is, when the balloons in the originating coils are at a maximum, those in the induced coils are at a minimum This is analogous to electrical *self induction*. A neighboring flexible exterior coil in the fluid is analogous to electrical *mutual induction*.

It is well to remember that pressure force applied to an area as is stress. Even though the applied force in the case of a hose is longitudinal, the reactive pressure is in all directions including radially to the hose. It is caused by the mutual repulsion of the electrons in atomic outer shells or energy levels. If the pressure is relieved the strain is relieved causing a counter pressure. When you turn off a water hose, you may notice the back pressure causing water to be forced out around the hose where the original pressure was applied via the shut-off valve.

This results from the energy stored in the neighboring ocean because of the stress on it from the fluid stream. This is analogous to the previously discussed back electromagnetic force that results when the voltage is removed from across an inductance that was attributed to the collapse of the magnetic field around the inductance.

Hydraulic Analogy for Magnetism

In the first edition of this book, I wrote: "magnetism has no direct hydraulic analogy." However, it seems plausible that when two streams of anything move in parallel to each other, the adjacent portions of the streams will be the portions that most affect each other. And if the streams are flowing in opposite directions it seems plausible that they will then bend each other one way in a third dimension whereas if they are flowing in the same direction they will bend each other the opposite way. That is, if the streams are moving in opposite directions their edges will not stop each other's flow as a whole, but it would appear plausible that the movement will be such as to bend the streams away from each other in the third dimension. And if the streams are moving in the same directions, the movement along their edges will not cause them to pull each other together as a whole but it would appear plausible that the movement will be such as to bend the streams in same directions with respect to each other in the third dimension. If similar reactions are demonstrated to be true in practice, they would certainly form a candidate for a hydraulic analogy to magnetism.

If this held up experimentally, two flexible-exterior hoses within a fluid with fluid flowing within them and so in the surrounding fluid [because of the ballooning effect pointed out earlier] would interact in a fashion analogous to that of magnetism between two parallel wires.

Setting up a physically confirming experiment could be quite cumbersome. Fortunately, this has already been done. In an effort by Dr Carl Anton Bjerknes in the Electrical Exhibition of Paris in 1881 to generally physically confirm hydraulic analogies to electricity., he demonstrated that pulsations in a fluid would occasion repulsion and attraction analogous to electrical magnetism. This obviously demonstrates our particular case for pulsating fluid filled flexible exterior hoses in a fluid.

Upon delving deeply into the World Wide Web, there is an article by Conrad W. Cook on the results of those demonstrations in which he wrote: "When two or more vibrating bodies are immersed in a fluid, they set up around them fields of vibration, and act and react upon one another in a manner <u>closely analogous to the action and reaction of magnets upon one another</u>, producing the phenomena of attraction and repulsion." He further wrote: "In this respect, however, the analogy appears to be inverse, repulsion being produced where, from the magnetic analogy, one would expect to find attraction, and vice versa..." The article does not go into his theoretical basis for claiming the observed hydraulic effect is the inverse of theory although one would suspect it referred to the, at least then, current mathematical formulation of electrical magnetism. I don't propose to delve into this further here. He further writes: "In this way Professor Bjerknes has been able to reproduce analogues of all the phenomena of magnetism and diamagnetism..."

Bjerknes's demonstration does not confirm the individual steps in my reasoning, but its results establish that there is a hydraulic analogy to electrical magnetism.

You may choose to visit giac2002.wordpress.com and under the BOOK LINKS heading, put your mouse pointer on the word "here" associated with link 14B, hold down the CTRL key and click to access more information on a hydraulic analogy to magnetism: "The-Hydrodynamic-Researches-Of-Professor-Bjerknes." [The referenced article, apparently no longer in publication, was in the Scientific American Reference Book: Manual for the Office, Household and Shop Unknown binding—1921"] Hydraulic analogy for magnetism.

Hydraulic Analogy for Mutual Inductance

If instead of another sheath-shell in the above illustration a water filled length of externally flexible hose were added, the water in that hose would follow the flow of the water in the original hose. That effect is called mutual inductance and the electromagnetic analogy is apparent.

Transformers

It's apparent that inductive coupling is proportional to the length in which the conductors are in proximity, the distance between conductors and the current flow or changes of current flow, Helical configurations of closely spaced wires increase the length within a given volume in proportion to the number of turns. This all leads to the coupling concept of ampere-turns.

If insulated wires are wound around each other and around a common air core multiple times, the effect will be multiplied in proportion to the ratio of the number of turns in the secondary coil to the number in the primary coil. The primary coil is the one to which an alternating voltage is applied, the secondary coil is the one into which an alternating current is induced.

The effect is even greater when helical windings are made around a common core of iron or some other ferromagnetic material in accord with the same principle involved with an electromagnet formed by winding insulated wire in a helical fashion around such a core while a steady current flows through the wire. In the latter case the resultant magnetic force increases in proportion to the number of turns around the helix as each turn is effectively another conductor so the magnetic effect of their combination as described between two parallel wires adds.

Hydraulic Analogy to Electromagnetic Fields

Again, consider a flexible-exterior hose filled with water in an ocean. If you apply a longitudinal pressure across it, the end on the side to which pressure is applied will balloon and compress the ocean water in the immediate vicinity and the fluid in the hose will start to flow longitudinally carrying the balloon along with it. That ballooning motion will be passed on to the ocean water immediately adjacent to the flexible-exterior hose. This process will continue through successive sheath-shells of ocean water around the hose although the amount of lateral water pressure due to successive balloon actions will decrease with the distance of each sheath-shell to the originating flexible-exterior hose.

Each sheath-shell will react backwards although with diminished radial pressure with radial distance from the hose. Finally, with time, for practical purposes the longitudinal motion in all sheath-shells and the hose will reach some equilibrium.

The inverse sequence of events will happen whenever the external longitudinal pressure is removed from the hose.

I believe the electrical analogy is obvious if pressure is replaced by voltage, the pipe by insulated wire and the ocean by electrical fields in empty space.

Hydraulic Analogy to Electromagnetic Wave Propagation

When the fluid in a flexible-exterior hose in an ocean moves back and forth, it will induce ocean throbbing at the same rate—as described in detail earlier. The throbbing amplitude will diminish with distance from that hose. The fluid in any other flexible-exterior hose in the ocean will flow back and forth at the same frequency as the ocean throbbing.

The remote flexible exterior hoses in such cases are analogous to electrical receiving antennas. The electrical analogy for the throbbing ocean is an oscillating free space electromagnetic field.

This is all further *beat-to-death* in the subsequent discussion of Maxwell's equations.

You would anticipate that a vertical flexible hose would radiate fluid omni-directionally in a horizontal direction. That is the same antenna pattern generated by a vertical electrical antenna—donut shaped with the center of the donut hole in line with the antenna axis.

Antennas of the same shape produce identical beam shapes whether transmitting or receiving. A parabolic antenna, for instance, is typically used with microwave fire control radars to produce a focused columnar beam for precision angular location of a target. A parabolic antenna used with sound, in any media including water or air, would also provide a means of focusing a transmitted or received sonic beam. There are many antenna types and arrays in use.

The linear dimensions of an antenna beam and so its generating antenna are proportional to the wave length in a particular medium. So the linear dimensions of sonic antennas and electromagnetic antennas with the same beam pattern differ considerably.

The speed of sound wave propagation in water is, dependent on temperature, around 5,000 feet per second. The frequency of a wave in a particular medium is the speed of wave propagation in that medium divided by the wave length. So a sonic wave length of one foot in water corresponds to around 5,000 Hertz—in the audio range. The speed of electronic wave propagation is roughly 200,000 times that of sound in water. So a wave length of one foot in a sonic medium corresponds to around 200,000 times 5,000 Hertz = 1 Gigahertz—in the lower electromagnetic microwave range.

Alternating Current Effects

If pressure is applied and then removed in a cyclical manner across the fluid contained in a flexible-exterior hose in a fluid ocean, the fluid flow in any other flexible-exterior hose in that ocean will be subjected to greater or less radial pressure in opposite phase to that in the original hose.

There will be initial impedance to current flow in such hoses due to induced water flow adjacent to them. The instantaneous impedance will lower as new fluid equilibriums are reached. Over time there will be an average reactance to back and forth fluid flow.
That impedance to a sinusoidally applied voltage across or current through an inductor is called inductive reactance.

Instantaneous inductive reactance is greatest when the sine wave is changing sign and least at the positive and negative peaks of the sine wave. Thus sinusoidal current flow through an inductor lags that through a resistor by 90 degrees and is 180 degrees out of phase to that through a capacitor.

Historically, electrical induction has been attributed to a separate field called a magnetic field where the magnetic field is always at right angles to the electric field.

Our electrical-hydraulic analogy points to the magnetic field being merely the effect of charge or molecular flow at right angle to the direction of flow.

The out of phase relationship caused by the ballooning interactions of parallel flexible-exterior hoses through which there is a back and forth fluid flow, where a maximum ballooning in the primary causing a maximum restriction in the secondary correlates exactly with the experimentally established Lenz's Law in electromagnetics: an induced electromotive force (emf) always gives rise to a current whose magnetic field opposes the original change in magnetic flux. The corresponding hydraulic analogy would be that a changing longitudinal pressure in the *primary* would change the progression of its horizontal ballooning action (analogous to a changing magnetic flux due to and electric primary) that would induce a counter-ballooning progression action in the secondary opposing the ballooning action progression in the primary.

All hoses have a hydraulic inductance and all lengths of conductors have an electrical inductance.

The schematic symbol for inductance is a horizontal looping representative of a spiral. This reflects the fact that a length of insulated wire wound in a helix presents a greater surface area per helix length than a straight wire of the same length and thus presents a greater area for inductive effects.

Tuned Circuits

To wrap up this overview of basic passive electronic principles, electric wave receivers can be tuned to a particular resonant frequency with a combination of capacitance and inductance. For instance, the combined impedance of an inductor and capacitor in series with each other will cancel at some particular frequency because inductive and capacitive reactances are 180 degrees out of phase with each other. So their combination would selectively let waves of that frequency pass through them hindered only by the unavoidable electrical resistance in series with each of them.

Electric Motors and Generators

After Oersted discovered that varying electric fields caused magnetism, Faraday, wondering whether the inverse effect were true—that is, would varying magnetic fields cause varying eletric fields, tested this concept experimentally and found it to be factually true. This very important fact ultimately resulted in the development of devices that would efficiently generate electricity—frequently simply called electric generators. Electric generators may be derived from the mechanical motion of conductors within magnetic fields. Inversely motion of electric motors is derived from the motion of magnetic fields. There is no direct analogy to electric motors and generators beyond a composite of the analogies to electricity and magnetism that this book separately discusses.

There are many forms of electric motors, both direct current and alternating varieties. In all cases their rotation is caused by rotating magnetic fields. In the direct current case, a magnet in a rotor is arranged with a magnet in the stator so the magnets first attract each other to align and then switch to repelling each other so as to cause continuous rotary motion. A direct current electric motor most simple in theory consists of an electromagnet stator close to but not touching a magnetic rotor so that applying voltage to the stator will align the opposite pole of the magnetic rotor with the stator magnetic pole, when the rotor just passes that aligned position the electric flow through the stator is reversed by a mechanical commutator so that its magnetic field is reversed and the rotor end is then repelled, etc.

Generators, dynamos or magnetos rather than converting electrical energy to mechanical energy as motors do, convert mechanical energy to electrical energy by mechanically moving magnets so their fields induce electrical voltages. The simple direct current electric motor just described could serve as a generator if the rotor were mechanically turned and the electrical connections used to supply electricity for the motor are instead used as pick-off contacts to supply electricity.

The combined facts of varying electric fields causing magnetic fields and varying magnetic fields causing varying electric fields were essential in the realization cemented by Maxwell's equations describing how one causing the other sequentially across space would result in a domino effect—the transmission of electromagnetic energy across space.

You may choose to visit giac2002.wordpress.com and under the BOOK LINKS heading, put your mouse pointer on the word "here" associated with link 18, hold down the CTRL key and click for information about generators.

Hydraulic Analogies to Individual Electrons

We have so far considered hydraulic analogies with respect to how currents behave in electronic circuits. Of course, wires do not physically balloon as we have suggested in the hydraulic analogy for electrical inductance. We do not see such any more than we see the electrons. Instead of physically, moving "balloons" the electrons in adjacent or nearby insulated wires distort space itself in a sort of milking action in the space between them and then in the wires. The hydraulic analogies to individual electrons are fluid molecules.

Active Electronic Devices

Active electronic devices are powered devices that control the amount of electric energy delivered to passive devices. These include vacuum tubes, also called electronic valves, and transistors.

Because of their much larger size and much lower reliability—basically due to their usual need of filaments to boil electrons off their cathodes—vacuum tubes have been displaced by transistors and other semiconductor devices in most cases except where high voltages are required, such as in transmitters.

In the case of a vacuum tube, the most basic electronic valve consists of a heated cathode constructed to emit electrons that are attracted to a positively charged anode after passing through a, typically negatively charged, grid with openings that unless charged would allow electron flow. The grid is used to control the number of electrons that flow through it by the amount of voltage applied to it. The electrons that reach the anode are then used to apply the desired voltage across passive elements.

A transistor consists of a semiconductor with its so-called base performing in an analogous fashion to the grid of a vacuum tube in that voltages applied to or current flowing through the base control the amount of current that will flow through the overall transistor from emitter through collector. The open loop, so-called gain, is frequently so high in transistors that it must be controlled with a feedback resistor. The earliest transistors were made of germanium and, so, too susceptible to temperature changes for military usage. The advent of silicon transistors overcame that difficulty.

Where vacuum tube design is typically done graphically via plate versus grid voltage curves supplied by manufacturers, transistor design is typically done with specified voltage or current ratios between elements. Transistors provide the capability of so-called *hole* [positive charge] flow rather than the exclusive electron flow in vacuum tubes. This provides some greater versatility in their use.

In both vacuum tubes and transistors, other than the three element devices are available and useful. Those include diodes with only a cathode and anode or an emitter and collector that simply block or pass electricity through them dependent on the positive or negative sign of the voltage applied across them.

Active electronic devices are used in transmitters and receivers, to simply suitably amplify voltages or current inputs or provide such as specifically tailored outputs of particular amplitudes and durations or derive suitable true-false logic outputs.

Ultimately, they all use or are limited by the passive elements we have described. Rather than the specifically tailored analog outputs for which they were originally used, they more frequently use digital inputs to process and provide required digital outputs.

Digital as well as analog logic circuits predated transistors and semiconductors, but the advent of the latter, led to mass production of electronic logic gates and specialized logic designers who did not necessarily require detailed knowledge of the electronics behind them. So I have chosen to continue discussion of logic design and logic design elements in a separate section: "Electronic logic design hardware and software."

Electrical Circuits

Electrical circuits include devices and the power sources designed to provide the required voltages or currents required by or to control the devices. Voltage sources are deemed to have the opposite sense, positive or negative, to the powered elements so that the total voltage around a closed circuit loop will be zero. [One of Kirchoff's laws.]

Wired circuits using transistors or previously developed semiconductor circuits or logic gates are used primarily only for developmental breadboards. The originally wired circuits are then replaced by printed circuit boards and may ultimately be encapsulated into new semiconductor circuits.

Logic designers are concerned with using logic gates—to be discussed separately—to derive desired objectives. Even they may be bypassed by general purpose computers controlled by programs designed to accomplish the desired ends.

Other Electricity Uses

Other well known uses of electricity include electric lights, heating elements, telephones and the array of electronic devices including televisions, radios and computers. Electric lights are most commonly made by heating a filament in an evacuated bulb to incandescence by passing an electric current through it. The light is caused by the filaments long-term burning or oxidation. There are of course many other ways to produce electric light, e.g. fluorescent and neon. Heating elements for whatever use, such as for comfort or in toasters are, of course, made by simply passing current through resisting conductors.

Overall Perspective on Electricity

Exploring the close analogies between hydraulics and electricity helps provide an intuitive understanding of electrical effects. We need to recognize that the effects themselves were individually discovered by experiments and observations.

The textbook *A First Course in String Theory*, Barton Zweibach, Cambridge University Press, 2004 summarizes this so well and succinctly that I quote it here: "In 1819 Oersted discovered that the electric current in a wire can deflect the needle of a compass placed nearby. Shortly thereafter Biot-Savart (1820) and Ampere (1820-1825) established the rules by which electric currents produce magnetic fields. A crucial step was taken by Michael Faraday (1831) who showed that changing magnetic fields generate electric fields. Equations that described all these results became available, but they were, in fact, inconsistent. It was James Clerk Maxwell (1865) who constructed a consistent set of equations by adding a new term to one of the equations. Not only did this term remove the inconsistencies, but it also resulted in the prediction of electromagnetic waves."

Hans Christian Oersted initially just happened to notice a deflection on a magnetic compass in proximity to an electric current carrying wire and then performed further precedent-setting experiments. We now recognize that the compass deflection was due to the electric current in the wire causing the wire to become magnetic and have learned that the cause of magnetism is always an electrical current at some level, macroscopically or at atomic and subatomic levels. But this wasn't known previously.

Faraday, established and used the concept of electric fields that now are believed to be distortions in empty space although in his era, even when the idea of fields became accepted in some quarters, there was a belief that they were generated in an invisible material called ether. Michelson-Morley in the early 20th century performed precision experiments that failed to show the existence of any aether and caused the idea of such an aether to be dismissed.

Scientists in general didn't initially accept the field concept. But Maxwell used it in developing his unified equations. He added a new term to existing equations *displacement current*, perhaps for mathematical elegance and consistency, that he postulated as being applicable to a dielectric and led to his unified "Maxwell's Equations." These when applied to free space suggested an electromagnetic wave with a speed determined by a combination of the previously experimentally determined constants of electrical permeability and permittivity of free space. Specifically the derived speed of an electromagnetic wave equals one divided by the square root of the product of the electrical permeability of free space times the electrical permittivity of free space [$v = 1/\sqrt{\mu\varepsilon}$]. That speed was consistent with the experimental determinations of the speed of light and suggested that light was one form of electromagnetic radiation.

In 1887, Henrich Hertz demonstrated such electromagnetic waves at what we now call radio frequencies. He did this with a simple transmitter that generated a spark through a gap connected to an induction coil and a simple receiver of looped wire and its own spark gap yards away that responded with its own spark. In other experiments Hertz also established that electrical conductors would reflect the waves and that they could be focused by concave reflectors.

It has been said that Faraday told Maxwell that he believed that the mathematics was an unnecessary frill that stood in the way of understanding. If true or not, such would be consistent with Faraday's intuitive way of thinking and mathematical education and knowledge, said to extend only to basic algebra. Faraday had intuited the possibility of electromagnetic waves and is said to have suggested such as an emergency-replacement lecturer to the Royal Society. But, of course, he was unable to quantify their speed of propagation without mathematics.

Maxwell closely studied Faraday's approach before formulating his equations and held Faraday in great esteem. There are those who believe the theory would have been better named the Maxwell-Faraday theory or vice versa. Maxwell would have likely agreed. Faraday had worked as a chemical laboratory assistant under Sir Humphrey Davy and even served temporarily as his valet. He was a commoner in England's then class-conscious society.

Davy is said to have revolutionized chemistry in his time. Faraday achieved his own renown as a chemist before becoming one of the leading historical figures in electrical experimentation and theory.

Maxwell's Equations: A Unified Theory of Electricity and Magnetism

Maxwell's Equations

I have been using three dimensional hydraulic analogies because I believe they more readily convey an intuitive understanding of electromagnetic phenomena which are three dimensional not including time than lines which are only two dimensional even when considered closed in circles. If centered at right angles to the longitudinal dimension of a linear antenna [or flexible-exterior hose], you might consider such circles as contour lines to describe a cylinder expanding in time within a fluid as a four dimensional model for an intuitive understanding of antenna radiation.

Historically, electricity and magnetism were separate areas of investigation for which Faraday used two dimensional representations via *lines of force* for each separate field showing how they interacted with each other. Maxwell based his mathematical equations on Faraday's *lines of force* concepts. So their discussion represents a different conceptual approach and their equations being mathematical require a temporary mental *shifting of gears*.

James Clerk Maxwell was one of the first to take Michael Faraday's use of fields to be more than useful ideas for theoretical work with electricity. After study and consultation with Faraday, he was able to consolidate most knowledge of electricity into four mathematical equations that unified Faraday's concepts of electric and magnetic fields to form one concise mathematical description of most of what was known about electro-magnetic phenomena. In addition to simply expressing observations mathematically, he added what is called a displacement current that proved all important in predicting electromagnetic wave propagation and will be further discussed below. Displacement current is simply that of a varying electric field in what appears to be empty space.

Maxwell captured in one set of equations most of what was known about electricity, magnetism and their interactions including their relationships in magnitude and time in the previously unknown electromagnetic wave propagation that they predicted. It is said that their original formulation included the Lorentz force on individual charges.

The Lorentz force is the combination of static electric and magnetic forces on charged objects. The static electrical force equals the product of the charge and the electric field and is collinear with the electrical field. The static magnetic force equals the product of the charge, the charge's velocity and the magnetic field at right angles to the direction of the velocity. The direction of this last perpendicularity as more precisely described as $\mathbf{F}_B = q\,\mathbf{v} \times \mathbf{B}$ and explained in the appendix with this combined equation for electrostatic and magnetostatic force: $\mathbf{F} = q(\mathbf{E} + \mathbf{v} \times \mathbf{B})$.

Maxwell's equations show that the magnetic and electric fields are perpendicular to each other in space and that the presence of a changing electric field in space is accompanied by a changing magnetic field in time and vice-versa, that is, a changing electric field in space is accompanied by a changing electric field in time. The equations show the precise orientation of conventionally assigned field lines.

The electric and magnetic waves may revolve around each other in what is called circular polarization, not revolve in what is called linear polarization, in a plane normal to the direction of propagation—vertical or horizontal—or in any way be perpendicular. Finally, they show that the divergence of the static magnetic field is zero while the divergence of the static electric field is proportional to any electric charge within the field, more specifically to the charge per unit volume within the electric field. These last refer to magnetostatics and electrostatics.

Divergence is the rate of change of a quantity such as a field—when what goes in one direction differs from what goes out in another.

Experimental observations actually show more than these bare mathematical equations indicate, they show that changing fields in space <u>induce</u> the changing fields in time—that is one precedes the other. This is important in understanding the back and forth action between electric and magnetic fields involved with the electromagnetic wave propagation in accord with the prediction of further mathematical manipulation of the equations. The equations mathematically describe the observed motions of electric charge that changes in the magnetic fields induce and the varying fields that moving charges produce.

The equation that shows the effect of a changing magnetic field in space not only shows that the field will induce an electric current in a conductor in that space that had been shown experimentally but that it will cause what has been called a displacement current in that space.

Maxwell added this mathematical term for displacement current that is crucial in the mathematical development showing the subsequently experimentally verified fact by Hienrich Hertz twenty some years later that electromagnetic waves propagate through space—via spark gap induced radio waves that exhibited interference, refraction and reflection. The addition of the term was suggested to Maxwell by the overall mathematical formulation as well a physically implied justification via capacitive action through a dielectric.

Displacement Current

Displacement current is what you would expect in a capacitor with no physical dielectric. Simply due to the known repulsive force between electrons without the need of a physical medium, the motion of electrons in a circuit connected to one plate of a capacitor will cause motion of electrons in the circuit connected to the opposite capacitor plate. This effect is evident across an air dielectric capacitor.

The concept of displacement current applies this to space in general over any distance. Even though, Maxwell may not have directly known that a capacitor with no physical dielectric whatsoever would perform the same way as an air dielectric capacitor, this seems plausible, and in fact is observably true. The concept of displacement current is crucial to the prediction of electromagnetic wave propagation.

Quoting from the Skilling text cited in the *References* section at the end of this book: "It was Maxwell who pointed out that a *displacement current*...would simplify and improve the mathematical system. Then Maxwell made a most remarkable proposal as follows: It is known by experiment that *conduction* current produces a magnetic field; *total* current is for mathematical purposes best expressed as the sum of conduction current and displacement current; is it not, then likely that a *displacement* current also provides a magnetic field?" I might say equivalently: "Is it not likely that an unobservable proposed current in space would generate a magnetic field as a *real* current is known to do?" Although experimental observations could not test the conclusions at the time. As we now know, they have been since confirmed.

It had been observed that a magnet must move with respect to a conductor to induce an electrical current in it. Given the concept of physical fields as alterations of the properties of empty space [or, then, of the *aether*] itself, this implied that a change in the magnetic field was required to cause an electric current in a conductor. That variation in the field could be thought of as a varying displacement current in the electromagnetic field.

In any event, Maxwell *stuck his neck out* for the time and added a mathematical expression for a displacement current that he symbolized as the rate of change of an electric field.

Maxwell's Equations Collectively

The crowning triumph of Maxwell's equations was their mathematical prediction that electromagnetic waves propagate through space at the speed of light, and so light itself was an electromagnetic wave as well as what became to be called radio waves, radar waves, etc. They mathematically captured almost everything learned world-wide since the pre-historic discovery of the lodestone about electricity, magnetism, their interactions and their relationship with each other in electromagnetic propagations in one set of equations.

You may choose to visit giac2002.wordpress.com and under the BOOK LINKS heading, put your mouse pointer on the word "here" associated with link 19, hold down the CTRL key and click to review a timeline of the history of electricity from 600 BC through 1999 CE noting the many people involved. Faraday is listed as beginning his electrical work in 1821, speculating that light might be electromagnetic with a transverse field vibration in 1846, retiring in 1855 and dying in 1867. Maxwell published a mechanical model for electromagnetic fields in 1861, stripping the mechanical component but leaving the equations, in an 1864 memoir and published his *Treatise on Electricity and Magnetism* in 1873 that was based on Faraday's viewpoint.

Although Maxwell's original formulations of his equations was likely in algebraic/trigonometric form, they can be much more succinctly described in the terms of vectors in association with terms of the calculus that we have discussed in those sections of this book. One further term is the operator called nabla or del that is a vector operator traditionally used to show a total derivative as derived from the partial derivatives of the vector fields—in rectangular coordinates, the partial derivatives of each of the three rectangular coordinate dimensions.

Regardless of how expressed, they summarize a majority of the known electromagnetic effects. Michael Faraday is said to have believed that the mathematical formulation *muddied the waters*. To some extent this is true, because it requires an understanding of the mathematical terms and operations involved apart from the physical observed results.

You can predict results without the need of knowing all the physical observations that lie behind the concise precise mathematical formulation of the results. The superlative book by Skilling, cited in the references section, relates the individual equations to the applicable observations and thought experiments.

The equations helped further establish the field concepts that have been extensively used by others including Einstein. Maxwell's equations with its use of vector concepts were invariant long before that property was accepted as being essential in such as the theory of general relativity in physics. Edward Purcell posed the question as to whether the ideas of relativity would have evolved "in the absence of a complete theory of the electromagnetic field" evoked by Maxwell.

Mathematics has an extremely important place in helping to ensure any lack of logical contradictions in physical hypotheses and theories and in practical usage in efficiently predicting results, but its expository usage for the general reader is dubious. Nevertheless some readers may be interested in the interpretation of each of Maxwell's equations shown in the appendix.

One of Maxwell's equations states that *a magnetic field changing in space will induce an electric current in a conductor* as had been shown experimentally. In addition to what I have just written the mathematics suggested the displacement current just discussed in addition to the current in a conductor. In any event, Maxwell *stuck his neck out* for the time and added a mathematical expression for a displacement current that he symbolized as the rate of change of an electric field. As with his other equations, that equation also shows a particular perpendicular relationship between the magnetic and electric fields.

In a medium without a conductor, his equations indicate that **a** *magnetic field changing in space will induce an electric field changing in time*. Another Maxwell equation indicates that *an electric field changing in space will induce a magnetic field changing in time*. In both cases, the word *induces* suggests that a change in one or the other field precedes the other in time and ultimately a *wave* motion propagating in space.

Again, mathematical solutions of the equations showed a speed for that wave which approximated the then known speed of light or any other electromagnetic wave. As discussed earlier, Henrich Hertz confirmed the existence of other electromagnetic waves than light in 1887, a little more than two decades after Maxwell's formulation.

Maxwell's other equations show that *there is no divergence in either field if the fields are acting in a media that does not contain a conductor or charge. But an electric field will diverge in a media with charge in proportion to the charge per unit volume within the media.* As you may be well aware, divergence is the opposite effect to convergence. These equations deal with electrostatic and magnetostatic field lines. The equation for the Lorentz force that includes that for magnetic force is included in the appendix

I generally prefer, as I believe most of you, not to dig into the mathematics except when having to to apply a principle. Part of the reason often may be that a reader may not know or remember the precise definition of mathematical symbols and concepts. To some extent these are touched upon in the chapter on mathematics.

But I suspect some of you would like to see each of Maxwell's equations explicitly addressed mathematically with the needed specialized terms explained. If you are one of those, refer to the appendix.

Directions of Field Lines and Return to Hydraulic Analogy

It is worth noting that the defined directions of the so-called magnetic lines of force are, perhaps confusingly, really at right angles to the direction that their *defined* force lines have been directed as opposed to the defined electric lines of force that are directed toward or away from the direction upon which the electric forces act. The so-called magnetic forces are always radial to a conductor and perhaps would be better named radial electric forces as you might naturally call their hydraulic counterparts—radial hydraulic forces. [Nomenclature is often confusingly chosen as in the case of conventional electric current direction which is exactly opposite to the usual mobile current carriers, electrons.] The hydraulic and electric radial and longitudinal forces are, of course, at right angles to each other.

Hydraulic analogs of Maxwell's equations would consider the electric field to be the hydraulic pressure in direct line with the applied force causing the pressure and the magnetic field to be the hydraulic pressure at right angles to the applied force. It is intuitively obvious that there will be an induced back and forth flow in the ocean that is analogous to free space.

The velocity of the electromagnetic wave derived from Maxwell's equations is one divided by the square root of the product of the permittivity [dielectric constant] times the permeability of free space whereas the velocity of hydraulic waves is the square root of the bulk modulus elasticity of the media divided by the density of the media or the square root of the change in pressure divided by the change in density. In the hydraulic analogy to wave propagation that we have set up, concentrations of fluid molecules moving longitudinally in a flexible-exterior hydraulic hose cause corresponding pressure variations in the just adjacent ocean fluid. These pressure variations cause current flow at right angles to the hose in correspondence with the molecules moving up and down in the flexible-exterior hose. These current flow changes will then cause corresponding pressure changes further yet from the hose. This sequence repeats indefinitely although magnitudes decreases in an inverse relationship with distance.

These show that the lateral electric field and magnetic field could be simply characterized as longitudinal and cross-wise manifestations of the same underlying field. Giving the perpendicular manifestations, different names ease mathematical manipulations. The historical reason for calling the lateral field a magnetic field is because the lateral field produced by a constant electrical current results in the macroscopic effects that we became first aware of in permanent magnets such as lode stones.

We have all seen expanding circle surface wave effects caused by the up and down motion of water. What we have just described shows that the same sort of thing happens below the surface. The analogy that we have made is between the actions and reactions in an ocean or any body of fluid with free space cyclically altered by a varying electric field.

These show that the lateral electric field and magnetic field could be simply characterized as longitudinal and cross-wise manifestations of the same underlying field. Giving the perpendicular manifestations, different names eases mathematical manipulations. The historical reason for calling the lateral field a magnetic field is because the lateral field produced by a constant electrical current results in the macroscopic effects that we became first aware of in permanent magnets such as lode stones.

Some might object to our hydraulic analogies as they have been historically dismissed historically to be replaced with field concepts. A hydraulic analogy may be reminiscent of a concept of an aether that was said to have outlandish properties [all pervasive, mass-less and infinitely elastic] and undetectable under the most precise investigation. We have since accepted the concept of fields in which free space itself can be distorted. Nevertheless it seems apparent to me that the hydraulic analogy with fluid molecules operating as if they are enormous electrons still provides, perhaps, the most outstanding vehicle to forward understanding.

On the other hand, some might object because of all the verbiage necessary to describe physical analogies as opposed to the brevity of mathematical expressions. But consider all the time in training and memory needed to even understand mathematical expressions not to mention their application to specific physical phenomena.

Antennas

Let's conclude the exposition of the way that electromagnetic waves propagate in free space without the need for remembering all the abbreviated mathematical symbology and operations: Concentrations of electrons moving longitudinally in an antenna induce corresponding displacement currents in free space, or any other dielectric, at right angles to the electrons moving up and down in the antenna. These induce corresponding electron field variations in a dielectric further displaced in the direction of wave propagation that in turn induce corresponding displacement currents at right angles to the direction of wave propagation. This sequence repeats indefinitely although magnitudes decrease in an inverse square relationship with distance.

As a sometime tutor, I have been aghast at a student picking up a calculator to find the product of eight times seven. On the other hand, it can be similarly disheartening to find that there are those who can hardly understand or communicate without the usage of mathematics. The last is not the case with the average reader.

We have focused on understanding of the basic principles of electricity here calling on hydraulic analogies to help. However don't be misled; there are many hydraulic properties and principles that have no analogies with those of electricity.

You may choose to visit giac2002.wordpress.com and under the BOOK LINKS heading, put your mouse pointer on the word "here" associated with link 20, hold down the CTRL key and click to view text associating these basics with many of the equations needed for practical applications.

Vectors and Mathematical Tools in Electricity other than basic Algebra and Calculus

Vector representations help illustrate the usefulness of mathematics in physical theories beyond the logical assurances that their long proven results offer. Mathematical forms are succinct as well as accurate. Humans have difficulty holding many things in their mind at one time. So long expressions and derivations are prone to error. Tensors, generalizations of vectors, provide an even more succinct symbology and procedures that led to Einstein using them in his development of the theory of general relativity.

Basic Alternating Current Circuit Equations

The basic mathematical equation for capacitive reactance is $X_C = 2\pi f C$ where X_C is the capacitive reactance, f is the frequency of the alternating current and C is the capacitance. 2π enters into the equations because 2π radians equals the 360 degrees involved in a complete cycle.

The corresponding equation for inductive reactance is $X_L = 2\pi f L$ where L is the inductance. The 2π in both equations is needed to account for the 2π radians to account for one complete alternating circuit cycle.

These combine in a way similar to that of parallel or series resistors in a circuit, as discussed earlier with respect to direct current, that is steady current, to obtain the total impedance with the resistors that are in a circuit—except allowance must be made for the fact that capacitive and inductive reactance are 180 degrees out of phase with each other and 90 degrees out of phase with resistance.

The basic mathematical relationship for current through a capacitor is $i = C \times dV/dt$ where i is the instantaneous current through a capacitor, C is the capacitance and dV/dt is the incremental voltage change with time across the capacitor.

The use of complex numbers where, for instance, resistance is treated as the real part and capacitive and inductive reactances as complex parts has proven so useful in electrical circuit calculations that their application is covered in almost any electrical engineering textbook.

References with Comments

Atkins, K. R, *Physics*, John Wiley & Sons, Inc. 1964. Extremely understandingly well written and understandable University of Pennsylvania general physics text book including mathematics with numerous illustrations.

Azimov, Isaac *Understanding Physics*, Dorset Press, 1960. This is Asimov's, always conversationally written, detailed overall physics exposition including three volumes within one cover. The volume specifically addressed to the subject matter of this book is *Light, Magnetism, and Electricity* is in its chapters, 12-Electromagnetism, 13-Alternating Current and 14-Electromagnetic Radiation. It avoids mathematics beyond citing some equations. To quote a note in its Electromagnetic Radiation chapter: "Unfortunately, they are differential equations, involving the concepts of calculus, and calculus is not used in this book. For this reason, Maxwell's equations will be spoken of, but will not be brought on stage." I write on the *concepts of calculus* which I believe are highly desirable, if not essential, for a better understanding of both Maxwell's equations and Newtonian physics.

Lieber, Lillian R., *The Einstein Theory of Relativity— A Trip to the Fourth Dimension*, Philadelphia Pennsylvania, edited and with a new foreword by David Derbes and Robert Janzen, Paul Dry Books, Inc. 2008 is a superb, more or less, popular book that describes tensor notation and operations as used in Einstein's theory of general relativity.

Purcell, Edward M. *Electricity and Magnetism*, McGraw-Hill, Inc. 1985 is an excellent exposition, especially on the theoretical basis for magnetism, by a Nobel prize winner in magnetism. This revision, Volume 2 of the Berkeley Physics Course, surprisingly for an elementary text in classical electromagnetism, required updates to account for then new developments in particle physics. Volume 2 is practically independent of the original volumes 1 through 3 because volume 2 had been extensively used alone. Its Appendix A, a concise review of the relations of special relativity, provides the absolute essentials of what was volume 1. Three other short appendices include—B on radiation by an accelerated charge, C on superconductivity and D on magnetic resonance. A substantially rewritten Chapter 4 includes sections on the physics of semiconductors.

Skilling, Hugh Hildreth, *Fundamentals of Electric Waves*, John Wiley & Sons, Inc, New York, Chapman and Hall, LTD, London 1948 is a superb textbook for a description of vectors and their use in conjunction with electric waves. A 1974 publication is available at Amazon.com. All reviewers, at this writing, give it five stars. One especially pertinent review follows: "Dr. Skilling shows exceptional insight into the physics of the electric waves. A must read for all who want a clear, concise understanding of electric waves."

Chapter 4
Feedback

Introduction

My interest in and comments on physics principle unification was largely around interactions that involve what appear to be actions at a distance (before the field concept in so-called empty space was accepted). Purcell offers a readily understandable everyday explanation of how the principles of magnetic induction follow from the augmented electric charge density caused by charge motion under special relativity. Ghosh offers an explanation of inertia based on his theory of gravitation that adds a velocity dependent term to Newton's theory in conjunction with which Mach's principle is exactly mathematically validated.

For now, I'm leaving my comments on general physics and mathematics principles to touch base on applied physics/engineering principles. Typical steps in an engineering approach include system analysis, development and test. Development generally involves use of established mathematical relationship with the known properties of available components, component purchase, system construction and some trial and error. Almost anything that can be conceived can be done; typical constraints are knowledge, safety, available dollars, available space and available time.

One of the more intriguing general engineering concepts is feedback—that is comparing an output with what the desired output is and automatically correcting the input to achieve that desired output. Unless otherwise modified, *feedback* in this book refers to *negative feedback*, that is feedback in which a sample of the output is subtracted from the input signal. Closely controlled positive feedback can improve signal amplification, but it can cause system instability to the point of causing oscillations. Acoustical regeneration caused by positive feedback to a microphone is an example of positive feedback.

Weather-vanes represent one example of built-in adaptive feedback in that the wind that the vane is designed to point towards automatically forces the vane to point towards it. The tail of an airplane uses the same principle in automatically enhancing flight stability by helping keep the airplane on course by using its own propulsion wind to do so.

All automatic feedback systems somehow sample the output and derive an error signal proportional to the difference between the output and the output desired to change the input in such a way as to achieve equilibrium at the desired output. Usually negative feedback is desired, that is the feedback acts to reduce the change that would otherwise occur in the output because of a change in the input.

Positive feedback to increase the resultant change in the output might be desirable in some circumstance such as gain control where some increased gain over *open loop* gain is desired. But positive feedback can easily lead to undesired instability such as oscillations.

When dealing with alternating current, negative feedback equates with out-of-phase voltages therefore if there are phase changing elements in the feedback loop, there may be some frequencies other than the primary operating frequency that become in phase because of such as temperature changes so that *parasitic* oscillations occur.

Elaborate control theory mathematics have evolved to achieve desired results while maintaining stability.

Servo-mechanisms

Perhaps the most common designed feedback in electro-mechanical systems are servomechanisms, also simply called servos. Perhaps the most prevalent provide the capability of remotely and stably controlling mechanisms in position. So-called synchros are an example.

Feedback Resistors

Feedback resistors are commonly used in electronic circuits to increase bandwidth and to ensure stability. They may be absolutely essential in the case of common emitter transistor circuits which may have enormous open-loop gain that can result in thermal runaway. One way or another feedback resistors use negative feedback to set the closed-loop gain at an acceptable level.

For instance, connecting a resistor to the collector in series with another resistor from the base to ground would form a voltage divider so that the amplification would become the ratio of the resistance of the resistor to ground to their combined resistance. If the open loop amplification were otherwise higher than this ratio, the signal at the collector would increase and feed an out of phase signal in this ratio to the base so that an equilibrium at the desired amplification would be established.

Digital Feedback

In digital circuits, the desired output digital number is determined by being directly inputted or from analog to digital conversion. That is subtracted from the actual digital output number and the difference used to drive the input towards a null. As examples:
1. The digital number corresponding to the radio frequency of an emitter can be determined by analog methods and subtracted from the digitally determined frequency of a jammer by the same method for digital to analog conversion control of the jammer frequency. Digital to analog conversion of that difference would then be used to tune the jammer frequency till the difference becomes zero.
2. The digital number corresponding to the angle of arrival of radio waves from an emitter can be determined by analog methods and subtracted from the digital number corresponding to the angular pointing angle of a jammer antenna. Digital to analog conversion of that difference would then be used to drive the jamming antenna till that difference becomes zero.

Fire Control Radar Range and Angle Tracking Feedback

The first exposure in my engineering experience in microwave radar based electronic warfare to major uses of feedback was with naval fire control radars—radars used for providing the *tracking angles* to a target for naval guns [on-board cannons]. In combat, a search radar [typically determining angle by rotating its antenna beam around 360 degrees and registering a radar return when the beam is pointed at a radar target] provides sufficient angle accuracy for someone to point the pencil-beam fire control radar antenna close enough to acquire the target—that is, *lock on* the fire control radar beam on the target. Typically a fire control radar operator would manually move the *range gate* to coincide with position of the radar target return shown as a *blip on* a display that shows the blip versus range from the ship [the time for the radar pulse from the ship to hit and reflect back to the fire control radar receiver]. The range gate would then automatically *lock on* to that range in conjunction with the fire control radar pencil beam.

Two feedback mechanisms would be typically used—one for range, one for radar antenna beam angle.

The visual *range gate* would coincide with an early and late electronic range gate so that the two would automatically move outward or inward in range to keep equal energy in both gates—that is, when more energy is detected in the late gate [farther from the ship's position], the range gates would be automatically moved closer in range or when more energy was detected in the early gate the range gates would be automatically moved further out in range. This establishes range tracking.

Of course the fire control radar pencil beam had to be directed on the target to make it possible for a range track to be established. If the target is a sea-surface target, one way to establish angle track is to sample radar return energy when the receiving antenna beam is switched slightly back and forth from port to starboard and automatically positioning the pair to maintain equal energy in both positions—that is when more energy is returned from the port beam, moving the antenna pair slightly starboard and vice-versa.

The transmitting fire control radar antenna beam is synchronized with the receiving beam pair, typically by having the transmitting and receiving antennas on the same antenna pedestal.

In general, feed back tracking systems require sampling outputs on both *sides* of a desired output value in an arrangement where the direction to reposition the *tracker* to achieve equilibrium (frequently to "drive toward the null") is contained in the output sample.

(In the more typical cases where shipboard fire control radars have to track other than surface targets, the analogous receiving antenna beams would circle the target, establishing a cone with respect to the line of sight between the target and the ship so that sinusoidal target return error signals sequences would be compared to the phase of the known sinusoidal beam pointing. The same basic principle as the antenna beam lobe-switching described above still applies to the receiving antenna beams that would, in these cases, drive towards the center-of-the-beams equilibrium position. Of course, rather than *lobe switching* or *conical scanning* of a received antenna beam from the same antenna system, appropriately relatively positioned simultaneous antennas beams can be used in analogous manners. These *monopulse radars* have advantages at greater cost.)

The same sort of target-tracking-feedback arrangements are used with so-called *active* guided missiles.

Chapter 5
Electronic Logic Design Hardware & Software

I hardly know where to begin—where we are in today's world or from basic electronic logic principles. Knowledge or arguably even understanding of basic electricity and electronics is not essential. However, digital electronic logic and computation is so obviously based on electronics today that some discussion of it can hardly be avoided. Read the *Digital Systems* text book reference by Tocci, et al, given at the end of this section if you are interested in more than the barest exposure to principles discussed here.

Very Basic Electronic Logic

Logic itself has, of course, always been an essential element of any design. Today, electronic logic design refers to the use of specially designed logic devices. Perhaps the most elemental electronic logic devices are logic gates that are used to electronically emulate the basic logic elements AND, OR and NOT. These elements were emulated indirectly or by name even in vacuum tube days, but now they are embodied in mass produced semiconductor devices.

Their basic use is in emulating Boolean algebra electronically. Boolean algebra was introduced by George Boole in 1854. It is a digital logic that requires only a limited number of logic relationships the result of which is either true or false. Typically, in modern digital logic and computer usage, a true result is signified by a 1 output, a false result by a 0 output. I will use that convention.

Simply, the logic relationship AND assigns a 1, *true*, only if all inputs are 1's, the logic relationship OR assigns a 1 if any input is a 1 and the logic relationship NOT assigns a 1 if the input is a 0 or it assigns a 0 if the input is a 1.

In electronic logic design and general purpose electronic digital computers, ones or zeros are differentiated by two different defined voltage levels, a one by at least a certain voltage and a zero by at most a certain voltage. Inputs could be set with such as physical switches, historically by punch cards, but now most likely by the action of electronic gates.

Electronic logic elements have been manufactured for use in logic design—including *and, or, not, nand* and *nor* gates. *Nand* gates for instance yield an output voltage indicating a one if any input voltage levels indicating a one are <u>not</u> set; *nor* gates yield an output voltage indicating a one if <u>none</u> of the input voltage levels indicating a one are set. Obviously hooking the output of a *nand* gate to a *not* gate will yield an output level from that gate indicating the same as a single *and* gate—that the inputs to both of the input levels to the *and* gate were set to a voltage level indicating a one. It is sometimes more advantageous to use *nand* or *nor* gates than the more obvious *and* or *or* gates.

Such gates are used for special logic design or in the design of the logic functions of general purpose digital computers. The basic if-then function that can be arrived at by a combination of these elements is subject to the same philosophical limitations of any crisp logic, due to its principle of the excluded middle. That is, given an input that is declared true any output of an if-then statement will be declared true. For instance, given the *if* input that the moon is made of green cheese then any *if-then* statement concerning Superman's birthplace will be declared true even if it is literally false. This is one origin of the computer usage truism—*garbage in, garbage out.*

The arithmetic functions of general purpose digital computers use binary numbers.

All numbers can be represented with respect to any base. Our everyday numbers use the base 10, Babylonians used the base sixty, binary numbers use the base 2. As with ordinary arithmetic, in learning to count in a base, a person or machine has the elements for computing in that base. In every case, the next count is one above the previous in the same place until reaching one less than the number base, then the next count is one in the next place to the left with all those in the previous place zeroed out.

Most simply in binary arithmetic, the sum of 1 plus 1 equals 10—that is a unit in the *two's* place and no unit in the *one's* place. In that system when one is added to 10 which represents two, a one goes in the next higher place and all lesser places become zero. That is 10 plus 1 equals 100 which represents the decimal system four in the binary system. [101 represents a decimal system five; 110, six; 111, seven; and 1000, eight.]

Of course, in decimal [base 10] arithmetic nine plus one becomes a one in the ten's place and zero in the one's place, etc. In octal [base8] arithmetic [an extension of binary frequently used in computers] eight plus one becomes a one in the eight's place and zero in the one's place, and sixty four plus one becomes a one in the 64's place with zeros in the two places further to the right.

The foregoing gives the flavor of the unique aspects of logic design. Logic designers become familiar with and use a vast array of ready-made circuit types that are also familiar to electronic circuit designers, but who traditionally design specialized versions for particular applications from the basic electronic circuit elements. An array of the generally available ready-made logic circuit types are described in terms of logic circuit elements in texts such as the one referenced here by Tocci, et al.

Computers

Special purpose analog computers using vacuum tubes dominated early electronic computer designs. Vacuum tubes were still used even when the less-prone-to-error digital approach became more appreciated with general purpose digital computers. Still vacuum tube computers required large rooms to house them and air conditioning to cool them because of their heated filaments. And like light bulbs, filaments would burn out resulting in unreliable operation.

Transistors were much smaller and required no filaments resolving those difficulties—even more so when it became possible to put circuits with many transistors in single semiconductor devices. Standardization and mass-production brought system prices down.

A special breed, logic designers, were spawned who needed to be familiar with Boolean algebra and at the equipment design level, with the variety of semiconductor logic components and circuits available as we just discussed. At the very beginning, computers were *programmed* with manual switches.

John Von Neumann is generally credited with the idea of including the particular program to operate upon binary and logical inputs to a general purpose digital computer within the computer itself because he wrote an original paper on it. [Although the architecture's description was based on the work of J. Presper Eckert and John William Mauchly, inventors of the ENIAC.]

Punched cards then magnetic and optical tape/disks became physical inputs for instructions and data for particular applications although they were programmed by software engineers. Languages for the engineers to use in preparing the software for computers evolved from instructions at the bit level, through basic computer instructions such as *fetch*, then so-called assembly languages and then more and more human oriented ones up to those that were user oriented.

Regardless of how input, the program as well as the particular logic or binary elements to manipulate is manipulated as programmed and the result put in the computer's memory. The computer's basic sequence once initiated is to access the memory location or locations given by an instruction to determine the previously set information at the memory locations, perform the indicated elemental operation, store the result in the indicated memory location and proceed to the next computer cycle.

As an example, a program might instruct the computer to access a given memory location for a step in a computation or logic sequence, provide values to use or name other memory locations for values to use with that operation and a memory location in which to store the result. The computer then would *fetch* the program instruction and values, perform the operation and store the result where instructed, Then proceed to the next assigned memory location for the next instruction, etc.
[e.g. Add, 2, 2, Store 4, Next instruction?]

When further instructed to do so, the computer would output its computed/logically-derived results to such as optical tape/disks for later use or even directly to a printer.

A basic computer operation can be completed in the order of millionths of a second or less so computers can perform operations very much more quickly than a human. It is generally more efficient to calculate the values in tables, for instance, such as those of trigonometric functions rather than retain the tables in computer memory. Also including extensive error checking that can be done at *light speed* puts computer output accuracies far beyond that typical for a human.

Human involvement remains essential to write computer programs as it is necessary to set up operations or mathematical equations to be solved. But computers will perform operations with very much less chance of error than a human doing so or correctly *cranking out* a mathematical equation.

Human memory was substantially re-enforced with the printing press and books. But once information is stored in computer memory, it can be much more rapidly accessed.

The knowledge that was once stored in human brains and books is now largely stored in computer networks and can be readily accessed by what has been known as googling.

Humans can still understand each other better and fill in the gaps that a question may leave open without the questioner having to guess the required effective literal follow-up questions. And humans can intuit the desirable follow-up operations when there is an unexpected computer output not covered by the original programming.

Prior programming to cover the universe of all possible human foibles or unexpected computer outputs is impossible as a matter of practice.

Reference

Tocci, Ronald J. Widmer, Neal; Moss, Greg *Digital Systems: Principles and Applications (11th Edition)*

I have never done detailed logic design, but I found earlier editions of this textbook helpful to understanding. Then it was attributed solely to Tocci. The following is a 2012 descriptive review among other laudatory reviews from Amazon.com buyers.

"5.0 out of 5 stars 11th Edition - A fantastic and concise textbook, April 27, 2012
By Ravindra V. Khire

"This review is from: *Digital Systems: Principles and Applications* (11th Edition) (Hardcover).

"If someone who is just starting to learn about digital systems or even someone who has years worth of experience in this field, asks me what's the best book in the market to get a thorough grip on the fundamentals of digital systems, this is the book. I may not have read every book, but I'll tell you this, it definitely won't get any better.

"It elucidates every point with numerous and well explained examples, from what binary numbers are to analog/digital conversion methods, memory, RAM structure, etc. It is worded in almost layman's terms so the essence is easy to pick up. Practical and relevant problems are given which further reinforce understanding. You also can't explain digital systems today without talking about VHDL and AHDL (Hardware Description Languages - HDL). Not only are the concepts explained through examples and diagrams, they're also covered by the HDL's, so if you're a college student where you'll most likely be introduced to them, this is ideal.

"I don't need to say anything more, as it's very clear how strongly I think of this book. Get it, it will make a huge difference in your understanding."

Chapter 6
Where to from Here and Now in Space & Time The Physics Odyssey

Based on the best available experimental evidence, light always moves through empty space at the same speed and it is not possible to observe any signal traveling faster than light. Everything we observe is approximately one nanosecond per foot in the past.

Edwin Hubble and Georges Lemaitre determined that, on average, the further stars are from us, the greater the apparent shift of their elements' spectra towards the red end of the spectrum. This *red shift* has been interpreted as a Doppler shift due to a general recession of the stars from us in space or the expansion of space itself. And that implies that the space of the universe is expanding as a whole. The constant of proportionality of the surmised distance and the red shift is called the Hubble constant.

Einstein had inserted a *fudge factor*, the so-called cosmological constant, in his original general theory of relativity to adjust its predictions that would have shown a dynamic universe as opposed to what was at that time believed to be a static one.

He later considered this as one of the *biggest blunders* in his life.

Because of the known finite speed of light, everything we see is in the past, it is just a matter of how far. We know that the light we see now from other stars is from their radiations up to billions of years in the past.

To avoid the unlikely assumption that the stars are receding from our particular sun alone, it has been speculated that the universe is similar to the interior or external surface of a sphere upon which the individual stars are scattered. Then from every star's perspective, the other stars appear to be, on average, receding as they would on the surface of an expanding balloon, representing expanding space. The further the stars are from each other on the surfaces, the more rapidly they would be diverging.

A basic question is how time and space are structured. Let's hypothesize that space-time as a whole is a four-dimensional hyper-sphere upon which every person's observations are as valid as everyone else's. We can accept that the hyper-sphere has mountains and valleys due to masses scattered throughout it in analogy to our earth's sphere. Although, not necessarily from a hyper-sphere point of view, this general idea, seems to fit the current viewpoint of physical scientists.

Newton's law of universal gravitation hypothesized that gravity works the same throughout the universe. Einstein's theory of general relativity is based on the same general idea as to gravitational fields. He sought, as perhaps everyone should, invariant measurements upon which all observers regardless of their position in space-time could agree.

We are free to hypothesize whatever we wish as long as the hypothesis does not contradict known observations or those that become known. Democritus, based on the very limited knowledge at the time, hypothesized atoms circa 400 B.C.—that is the idea that all that exists is atoms and the void. The idea was not accepted for millennia. However it is a standard hypothesis now in modern physical science—expanded from the idea of atoms to other than the molecules that Democritus seemed to envision to electrons and particles of sub-atomic size.

Einstein's theory of relativity, Newton's laws and Maxwell's equations concern interactions with no intervening observed matter. This concerned scientists to the extent that they had postulated an a*ether* that required very strange if not contradictory properties.

The precisely measured Michelson-Morley experiments in attempting to determine variations in the speed of light through a*ether* completely failed to show any shifts. This not only led to Einstein's postulate that the speed of light was constant as measured in any frames that were moving at a constant velocity _relative_ to each other, it dispatched belief in an aether. Einstein would have preferred that his theories were not called relative but theories of invariance because they showed that there were invariant relationships that would be measured the same in any frame of reference.

His theory of general relativity rested on finding such invariants in four dimensional space, three in space, as we see it and one of time, and his principle of equivalence [that inertia and gravity are equivalent].

Of course, in science, any hypothesis needs to stand up to future observations and measurements.

Apparently Einstein's theory remains to prove remarkably accurate.

You may choose to visit giac2002.wordpress.com and under the BOOK LINKS heading, put your mouse pointer on the word "here" associated with link 21, hold down the CTRL key and click to read of an instance corroborating this. It obviously forgoes the need of any type of material *aether* for so-called action-at-a-distance. The idea of such instantaneous action has been discarded.

You may also choose to visit giac2002.wordpress.com and under the BOOK LINKS heading, put your mouse pointer on the word "here" associated with link 21A hold down the CTRL key and click for more information on detected gravitational waves.

Newton's laws assume an absolute space with so-called fictitious or inertial centrifugal and Coriolis forces with no apparent sources but which need to be invoked for his laws of motion to work. The idea of an absolute space worried Newton although he had no idea what to do about it. Nowadays space itself is assumed to be deformed by mass.

Centrifugal effects occur whenever an object is turning with respect to what Newton called, regretting the need to do so, absolute space. They are evident wherever an object is turned from a straight line (or geodesic). The water in a bucket will stay in it when it is whirled around at sufficient speed. You could determine the amount of absolute rotation of a body containing a liquid by measuring the curvature of the liquid surface rotating with the body. The earth itself is flattened at the poles and bulging at the equator in apparent response to a centrifugal force *caused* by the earth's accelerated motion with respect to *absolute space*. Centrifugal effects, as the name suggests, seem to be caused by a force directed outward from the center of rotation.

Coriolis effects occur on a uniformly rotating frame of reference such as the earth. The Foucault pendulum hanging from a universal joint so it can swing freely in any direction will change its apparent direction of swing around it as the earth actually turns around it every 24 hours. Other indications that it is the earth and not the heavens that rotate every 24 hours include gyroscopes which seek to maintain a constant orientation to the so-called absolute space and that an object dropped from a height will land to the east rather than directly vertically below from where it is dropped. Coriolis effects appear to be forces at right angles to the axis of rotation and to the directional velocity of a body.

From these effects, we can see that we can determine the absolute acceleration of a body, although we can not directly determine any absolute velocity. Bishop Berkeley, followed by Ernst Mach putting a finer point on it, observed that these *accelerated* motions attributed to an absolute space, were *relative* to the fixed stars. Einstein's principle of equivalence between inertia and gravity strongly suggests that inertial effects are due to the aggregate mass of the distant fixed stars. Sciama argues that the aggregate mass of those shells of matter would be infinite, as would the brightness of the night sky, were it not for the expansion of space evidenced by the red-shift of the spectra of their contained matter.

It makes sense that the aggregate gravity of distant matter on all sides of any body would resist any accelerated motion of the body—its inertia. It's apparent that a body at rest in such an environment would be in such equilibrium that it would have no net force upon it to leave that state of rest. It is not so intuitively evident that a body in steady motion would not have a net force acting upon it. However after a body is in motion in a vacuum, it has become in equilibrium in the gravito-inertial field, so why should it not continue so?

If you consider the surrounding gravito-inertial fields set up in ancient times by the ancient distant stars, you would think that any displacement or velocity once established is in balance with respect to long-established fields pulling in opposite directions equi-distant from an object at rest at any displacement due to past motion. The established gravito-inertial field was caused by the conglomerate distant matter, dark matter and dark energy.

This is not subject to experimental diddling even if the distant mass could be changed because its effect would be delayed by light years. The only diddling known possible is the introduction of closer mass (which is already accounted for by both Newton's gravitational theory and Einstein's General Relativity [EGR]). Presumably inclusion of gravity terms in the EGR of an amount corresponding to the distant aggregate mass in all directions from an object, i.e. an overall curvature, would yield the same inertial effects including centrifugal and Coriolis effects as experienced in reality. If not, and the fact that there is no consensus among cosmologists on Mach's principle, indicate that such EGR adjustments have not been found and that general EGR alone is not quite adequate and needs tweaking.

A conclusion is that your choice of either the concept of absolute space or the aggregate mass of distant stars, molecules etc. could yield the observed inertial effects. The question is: "Is there a mathematical model or one yet to be developed that would completely explain current observations and perhaps predict further observable results that are not yet known?"

Cosmologists continue to advance modified gravitational hypotheses. You could assume that any that might prove to be successful have to be in agreement with Einstein's theory of General Relativiy [EGR] as far as it goes, in the way that EGR is in agreement with Newton's theory that still prevails as an approximation under conditions of relatively low gravities and low relative velocities.

Many sources seem to me to say that the Einstein theory of general relativity as applied to gravity and quantum theory are incompatible. I can claim no expertise in this area. However a wiki article on quantum gravity says they are compatible at low energy levels. You may choose to view that article on the Book Links page of giac2002.wordpress.com by putting your mouse pointer on the word "here" associated with Link 30A, holding down the CONTROL key and clicking. This selection on quantum gravity includes the following quotation: "While confirming that quantum mechanics and gravity are indeed consistent at reasonable energies, it is clear that near or above the fundamental cutoff of our effective quantum theory of gravity (the cutoff is generally assumed to be of the order of the Planck scale), a new model of nature will be needed. Specifically, the problem of combining quantum mechanics and gravity becomes an issue only at very high energies, and may well require a totally new kind of model."

You can view an article, that was "substantively revised on May 27, 2015," by clicking Link 30B using the same procedure described for 30A above. It shows the limits of overall knowledge on quantum gravity theory: "Quantum gravity is beset by a combination of formal, experimental, and conceptual difficulties."

Stephen Hawking has a more positive outlook than that expressed in the above quotation from the Stanford Encyclopedia of Philosophy. With regard to perhaps the most advanced superstring theory that sprung from string theory and super-symmetry, M-theory: 'M-theory is the only candidate for a complete theory of the universe."— mainly due to lack of viable alternatives.

Wikipedia defines string theory "as a theoretical framework in which the point-like particles of particle physics are replaced by one dimensional objects called strings. It describes how these particles propagate through space and interact with each other." This has been expanded into what has been called M-theory. A further discussion of M-theory is found in the Wikipedia article: "Introduction to M-theory" that strings are really one-dimensional slices of a two-dimensional membrane vibratinng in 11 dimenstional space-time. You can view the Wikipedia article "Introduction to M-theory" on giac2002.wordpress.com by putting your mouse pointer on the word "here" associated with Linck 30C on the Book Links page holding down the CONTROL key and clicking.

The dimensions are said to be "folded within" the four dimensions of space-time that we traditionally perceive or use. These dimensions are said to be so tiny that they are not even remotely close to anything that can be observed with current instrumentation. One might be tempted to ignore a theory which, accordingly, is so, not falsifiable by observation. On the other hand, many particles, electrons to name some, are not directly observable but are generally accepted as real by physicists because of observed effects that they are deemed to produce. On the other hand, many particles, electrons to name some, are not directly observable but are generally accepted as real by physicists because of observed effects that they are deemed to produce.

"Some phusicists are skeptical that this approach will ever lead to a physical theory describing our real world due to fundamental issues."... "Some cosmologists are drawn to string theory because of its mathematical elegance and relative simplicity... Stephen Hawking has stated: "M-theory is the only candidate for a complete theory of he universe"— mainly due to lack of viable alternatives.

The theory ascribes the known physical particles to various vibrations of its strings, more recently, *superstrings*, and gravitons appear to just "fall out" from it. Apparently there is no other known theory from which gravitons, that is gravity quanta are derivable. On the other hand strings are much smaller than there is any current capability to detect. This last does not scuttle a theory even though the physical detection of strings *would* enhance their credibility.

A bottom line in all of this is that although we have learned enough to make much more accurate predictions in time and space in the last century or so, there still remain unknowns to explore with theories to account for them. Explanations of cosmological observations have led to alternate hypotheses for theories of gravity.

These include hypothesized scalar-tensor theories including Brans-Dicke and those to which D.W. Sciama has contributed including an Einstein-Cartan gravity also known as the Einstein-Cartan-Sciama-Kibble theory. The latter was published several decades later than Sciama's *The Unity of the Universe* and *The Physical Foundations of General Relativity*.

Sciama and others have included considerations of *dark matter* that is unobservable but appears to be considerably more abundant than observable matter.

Some say that the currently observed effects sometimes attributed to the presence of dark matter and dark energy and acceleration of its expansion might conceivably have changed Einstein's later views leading to his abandonment of his earlier views and attempts trying to incorporate Mach's principle that inertia was caused by the mass of the entire universe. You may choose to visit giac2002.wordpress.com and under the BOOK LINKS heading, put your mouse pointer on the word "here" associated with link 32, hold down the CTRL key and click for information on dark matter. You may also choose to visit giac2002.wordpress.com and under the BOOK LINKS heading, put your mouse pointer on the word "here" associated with link 33, hold down the CTRL key and click for information on dark energy.

One of the ideas that may have contributed to Einstein's abandonment of his earlier views and attempts to incorporate Mach's principle into his general theory of relativity was the theoretical de Sitter universe that *without any mass* still incorporated inertia in that universe while still in accordance with Einstein's original theory of general relativity. You may choose to visit giac2002.wordpress.com and under the BOOK LINKS heading, put your mouse pointer on the word "here" associated with link 31, hold down the CTRL key and click for information on a De Sitter universe.

Some sources say that the currently believed acceleration of the expansion of the universe might lead to resurrection of Einstein's cosmological constant that he had inserted in his original general theory of relativity so it would accord with the view of all scientists at the time that the size of the universe was constant. Einstein abandoned it after Edwin Hubble found that the universe was expanding at a constant rate and subsequently said his introduction of the cosmological constant was the greatest mistake of his life.

Remember light was early-on considered to be composed of particles, then supplanted by the conviction of many if not most scientists that it was strictly a wave phenomena then, a photon particle, and now it is currently, accepted to have both particle-like and wave-like properties. Even the most widely accepted scientific *truths* are subject to alterations based upon further observations.

Einstein's Theory of General Relativity is assumed as part of many hypotheses. Other hypothesized theories include an 11 dimensional superstring, *M-Theory*. The latter is even more ambitious in that it *aims* for a theory of everything, that is, one that accounts for relativity, electromagnetism and quantum mechanics.

One especially compelling theory postulating Mach's principle that does not incorporate Einstein's Theory or, for that matter, deny it, is one advanced by Dr. Amitabha Ghosh..

Chapter 7
Unified Theory of Gravity and Inertia

Amitabha Ghosh's book describing his theory in detail: *The Origin of Inertia—Extended Mach's Principle and Cosmological Consequences* begins by showing the difficulties encountered with the current theories including but not limited to those previously pointed out by Sciama in his attempt to quantify Mach's principle—that inertia is due to motion with respect to the fixed stars. Ghosh also points out deficiencies in other proposed hypotheses. He asserts: "Einstein developed his General Theory of Relativity with the aim of incorporating Mach's Principle. Unfortunately he did not succeed."

Ghosh's book notes that by using Hubble's constant and the speed of light Sciama estimated the size of the observable universe and that this with an estimate of the mass density of the universe arrived at an impressive approximation of the equality of gravitational and inertial mass. To achieve *exact equivalence* with this method you would need an extreme fine tuning of the mass density and the size of the universe.

Einstein had advanced the principle of the equivalence of gravity and inertia. And precise experiments have corroborated the equivalence of the gravitational mass and inertial mass of matter. In conjunction with Ghosh's theory, Mach's principle is shown to result in just such an *exact* equality of gravitational and inertial mass.

Ghosh's theory of gravito-inertia follows from his adding a closely reasoned velocity-dependent term to the static term in Newton's law of gravity in addition to the acceleration-dependent term proposed by D. W. Sciama. The added velocity-dependent term results in expressions accounting for an estimated aggregate universal mass mathematically canceling out so that the calculated inertia due to acceleration does not rely on an aggregated universal mass estimate. It also results in a calculated universal drag due to velocity.

Added comment by Amitabha Ghosh: "Though the quantitative value of the 'velocity dependent inertial induction' term is very small it has a very important role. This term leads to a 'drag' force that acts on everything including gravitons. This in effect produces a feedback effect that makes the equivalence exact. This is not possible otherwise. Thus the philosophically unacceptable extreme fine tuning is avoided.

Another important point of the 'velocity dependent' term is that it has to be not just velocity dependent but must be a resistive force (that is it must oppose the velocity). Just adding a term whose magnitude depends on velocity will not yield the results. Quite often this point is missed."

This velocity drag is largely experimentally imperceptible and had not been overtly investigated because it was previously unknown and unexpected. But it does yield explanations for some previously observed unexplained phenomena and alternate explanations for other, sometimes ad hoc, explanations.

Ghosh's calculated inertia from acceleration is exactly the observationally established inertial relationship—the gravitational mass of an isolated body times its acceleration. This process mathematically validates Mach's principle when you recognize Sciama's mathematical established point that the aggregate cosmological gravitational force and so universal inertial field of far distant masses far outweighs the contributions of nearer masses.

Einstein as many thinkers including Bishop George Berkeley, Gottfried Wilhelm Leibniz and of course, Ernst Mach found Mach's principle compelling and hoped to confirm it with his general theory of relativity but was unable to do so. The foregoing result of Ghosh's theory recommends its general acceptance by physical scientists.

Ghosh's further mathematical analyses of his unified theory predict physical phenomena not previously established and provides alternate explanations for many physical phenomena that had previously required ad hoc explanations. Unfortunately, the effect of his added velocity-dependent term is generally so small that separating it from other factors is most frequently beyond current experimental capabilities.

His added velocity term is proportional to the square of an object's velocity divided by the speed of light so it could be expected to be more evident at high speeds. His theory has potential far-reaching consequences in cosmological theory. Ghosh's further mathematical analyses examine some consequences that are discussed below.

The *cosmic drag* due to Ghosh's added term modifies Newton's first law of motion that states that an object in motion will forever remain in its original state of straight-line motion unless acted upon by a force. Ghosh's theory predicts, for instance, that all rotation slows down regardless of how generally imperceptible.

Added comment by Amitabha Ghosh: "The earth has a large enough moon that it can take some part of angular momentum of the earth (as in the standard theory) and still be near the Earth in spite of the increased orbital angular momentum. On the other hand the planet Mars has two extremely small satellites—Phobos and

Deimos. They cannot take any perceptible amount of Mars' angular momentum. The Extended Mach's Principle provides a mechanism for angular momentum transfer without any contact and predicts a secular retardation of Mars's spin due to its interaction with the Sun. I approached NASA in 2005 for measuring Mars' spin. NASA is sending a spacecraft that will land on Mars in 2016 and one of the major goals project is to measure Mars' spin very accurately. If the predicted retardation is found to exist it will be difficult to explain it through conventional mechanics."

This provides an explanation for the established observations that the rotation of the earth and the speed of the revolution of the moon around the earth for instances have slowed over the millennia. Other ad hoc explanations had been advanced for these phenomena of course.

Ghosh's-theory cosmic drag also provides alternate explanations for energy retardation including the observed cosmic redshift of the spectral lines of elements with distance that corresponds to those discovered by Hubble. In fact, Ghosh's calculations show a Hubble constant relationship between redshift and distance quite close to observational estimates. His theory also predicts redshifts near masses that correspond to the time retardations shown with Einstein's theory of general relativity that has been experimentally established.

The cause of the Hubble constant redshift is really still not resolved. If Ghosh's theory were available at the time of Hubble's discovery, the Doppler effect would not have had to be called upon as the only known potential cause of the red shift. The presumed Doppler cause was the initial impetus for the big bang theory.

Ghosh's theory provides a basis for a quantitative fall off of gravitational force with extreme distance due to cosmic drag [using Sciama's estimate for the average mass density of the universe, Ghosh estimates: "To get a 50% drop in the value of G we have to go very far-about 10 Billion light years!"]. That would resolve the paradox that otherwise would result with an on-average homogeneous mass in an infinite universe. The hitherto presumed Doppler cause of the redshift does the same qualitatively as pointed out by Sciama for that paradox as well as Obler's paradox which would otherwise make the night sky as light as day.

Ghosh's theory predicts a directional variation in the average recessional redshift dependence on variations in nearer aggregated mass. This has been observed.

Ghosh has mathematically shown instances where the results according to his theory are much closer to observational results than the conventional theory. These include the redshifts occasioned by white dwarfs and the limb of the sun. Ghosh's theory also shows redshifts that are not explained by conventional theory such as those observed when photons graze a massive object.

Ghosh's theory mathematically predicts that a rotating ring, spherical shell or sphere of mass will induce a force on a particle that is not predicted by conventional physics. So a rotating planet will occasion a transfer of angular momentum without physical contact. He writes in his book: "no similar mechanism exists in conventional physics."

It also predicts a "torque on a rotating sphere in the vicinity of a large massive body." Applying this or other explanations for the known retardation of the earth's rotation is problematical because of what appear to be currently insurmountable difficulties. "This is reflected in the comments of two noted experts in the subject, Calame and Mullholland..."We must abandon the habit of treating the unexplained acceleration as being entirely of tidal origin and search for other causes that might contribute to it."

At the end of a separate chapter of extensive mathematical analyses in his book, Ghosh summarizes: "[A] number of different cases has been considered where object to object velocity-dependent inertial induction can have detectable effects. It is clearly established that in all these cases the observational results strongly support the proposed theory. In most cases, the model resolves the unsolved mysteries and unexplained features."

Each body of mass in a Ghosh universe could be conceived as enmeshed in all the rest of the mass in the universe with the universe's composite mass attracting it in a state of equilibrium unless acted upon by nearer forces. In such a gravitational force concept the cosmological drag *tired-light* velocity term could account for all the redshift.

It is not evident that Ghosh's approach contradicts Einstein's theory of general relativity in which detailed gravitational effects are given as caused by curvature of the time-space continuum rather than forces-at-a-distance between masses. If anything, Ghosh's approach would appear to only affect the Einstein theories required gravitational field inputs perhaps with an overall curvature decreasing at extreme distances as Ghosh's theory indicates.

It is worth noting, that the universal gravito-inertial field postulated in all hypotheses citing gravity to account for Mach's principle, combines the effects of mass not only far away in space but ancient, some, millions of millennia in the past. A field's strength is modified by the distance to gravitational sources, but has no bearing on action-at-a-distance counter-arguments which claim such theories as Ghosh's presuppose instantaneous cosmological effects.

The Ghosh theory deserves and hopefully will receive serious precision experimental efforts to further test it. You may choose to visit giac2002.wordpress.com and under the BOOK LINKS heading, put your mouse pointer on the word "here" associated with link 22, hold down the CTRL key and click for.an outstanding further review of Ghosh's book.

References with comments

>Ghosh, Amitabha* *The Origin of Inertia*, Montreal, Quebec, Apieron, 2000 [(157 pages) clearly written, although using mathematics extensively to the level of calculus, without the need of invoking tensor analysis.]
>
>Sciama, D. W. *The Physical Foundations of General Relativity*, New York: Doubleday,1969 [Short (104 pages) and clearly written largely non-mathematical book, based on the physical and conceptual foundations of General Relativity]
>
>Sciama, D. W. *The Unity of the Universe*, Garden City New York, Doubleday & Company, Inc. 1961

**Amitabha Ghosh served IIT Kharagpur from 1997 to 2002 as the Director and subsequently returned to his Professorial position at IIT Kanpur where he continued till his retirement in 2006. After retirement he became Platinum Jubilee senior Scientist of The National Academy of Sciences, India and Honorary Distinguished Professor Aerospace Engineering and Applied Mechanics Department Bengal Engineering & Science University, Shibpur Howrah, India 711103*

Chapter 8
Putting It All Together on Gravity & Inertia

Putting it all Together

Objects not only resist changes in motion, that is changes in an object's speed, they also resist changes in the direction of that motion even though that change in direction does not appear to be relative to any other specific body of matter.

An example is the Foucault pendulum, a pendulum that is suspended by a universal joint that allows it to not only swing back and forth but for the plane of its back and forth motion to rotate. Rotate it does in synchronism with the earth's rotation with respect to what could be called some absolute space. That is, it appears to rotate with respect to the earth. The earth itself is flattened at the poles and bulging at its equator. Again this is taken as because the earth revolves with respect to something around the earth which could be called absolute space.

Newton in his *laws* of motion and gravity accounted for observed inertial phenomena, by assuming that motion was relative to an absolute space. The philosopher, George Berkeley, noted that such motion was relative to the *fixed* stars. Ernst Mach took that one step further and proposed the explanation that inertia was caused by the total mass of the universe.

Another way of totally describing inertial effects is to observe that the motion of objects resist any departure from following a straight line. That raises the question as to what determines a straight line. For instance, how does light know which direction to go? Einstein's theory of general relativity is based on the idea that a straight line, more generally called a geodesic, is determined by the so-called curvature of space, or lack of it, that is determined by the gravitational mass due to the matter and energy in that space. In other words, it depends on the composite gravitational field.

Mach's principle or some other explanation is required to adequately explain all the inertial observations described above. It is unclear that Einstein successfully incorporated Mach's concept that inertia is caused by the total mass in the universe into his general theory of relativity. Because of this or other perceived inadequacies, there are several proposals to augment his theory or the gravitational inputs required for solutions.

Nevertheless, Einstein enshrined a *principle of equivalence* between gravitational mass and inertial mass in his development of general relativity. And the best available experimental observations indicates an exact equivalence between gravitational mass and inertial mass.

Dennis Sciama was able to show an approximate correspondence followed between inertial and gravitational mass with a speculation on the total mass in the universe based upon observations.

Amitabha Ghosh has been able to show the exact observed equivalence of inertial and gravitational mass via the mathematical development of his unified theory in which the exact mass of the universe mathematically cancels out.

Understandably, the question may arise as to how any theory and Mach's conjecture that depend on the aggregate mass density of the universe can rule out the need for any consideration of the absolute value of that aggregate mass density of the universe. The answer is that Ghosh's unquestioned mathematical derivation of this follow inexorably from his plausible mathematically precise defined postulate for his added velocity component to gravity.

His theory is, so far, physically established to the extent that it offers explanations of many effects that previously had nothing but ad hoc explanations tailored to specific individual effects. His theory also explains other effects that have not yet been observationally verified [or falsified].

As with all theories, further observations may determine which of all the competing theories best describe reality. It may be that the Ghosh's mathematically consistent theory that extends Newton's *laws* will stand alone without the need of Einstein's theory of general relativity. Or it may provide another gravitational input to EGR that will enable the combination to cement the validation of Mach's principle via the concept of absolute space being replaced by that of a universal gravito-inertial field.

This leads to a discussion of the application of Einstein's theory of general relativity to being in order with a few further clarifying details.

Application of Einstein's Theory of General Relativity

The mathematics behind Einstein's general theory of relativity consists of tensor equations that represent the curvature of four dimensional space-time corresponding to the physical observations of the gravitational effects as embodied in Newton's *law* of gravity. [Einstein had to make a choice of potential solutions based upon what had been physically observed.] Specific exact solutions are hard to come by because of the mathematical complexity of the equations and their non-linearity—occasioned by the fact that the fields they describe, themselves exert gravitational forces.

When Einstein originally proposed them, he was only able to approximate their solution using the then experimentally derived values of gravitational forces. He was surprised when Karl Schwarzchild obtained an *exact* solution of Einstein's field equations within little more than a month after their publication—apparently based on a fine tuning of the physical situation abstraction by specifying a spherical, uncharged non-rotating mass as the gravitational source. Schwarzchild's results were the same that Einstein had obtained with mathematical approximations.

The scientific world remains enamored with the beauty of Einstein's theory of general relativity and, after a century of continued physical observations its results have not been proven invalid.

On the other hand, although Einstein was enthralled with the idea of Mach's principle, he was apparently unable to successfully incorporate it into his theories of relativity. Mach's principle was so intriguing because it suggested an explanation of physical phenomena for which Newton had thrown up his hands and unhappily postulated an absolute space—phenomena such as the Foucault pendulum and centrifugal effects. Two isolated explanations for gravity and inertia has been considered one explanation too many.

One expression of Mach's principle was that such effects were caused by interactions with distant stars or all the composite mass of the universe. Sciama in his book: *Unity of the Universe* made compelling arguments for the interactions being caused by distant stars—primarily based upon the aggregation of mass from an essentially homogeneously distributed universe of stars [as seen in the observable universe] in their far distant spherical shells based upon the geometric truism that the surface of a sphere is proportional to the square of the sphere's radius.

Sciama noted that such an argument would mean that the aggregate starlight at night would exceed that of the noonday sun [the Obler paradox] and that if that had been taken into account, the expansion of the universe would have been realized long before Hubble's observations with the Doppler Effect. *Tired light*, as with one aspect of the Ghosh theory, would explain Hubble's observations along with the Obler paradox without the need of the expansion of the universe implied by a Doppler effect.

Chapter Eight

Conclusion

Einstein's basic Theory of Special Relativity postulates seem to be overwhelmingly backed by all observations so far made. Given these postulates, its results are mathematically inevitable. And as I understand it, his Theory of General Relativity provided a set of tensors that, being tensors, ensure the desired invariance of Einstein's conjecture as to the curvature of space given gravity value inputs and the correspondence of the results with then known observations.

His equations have not since been invalidated by a century of further observations, but that does not necessarily preclude the equations being further refined such as by further observations or gravity input refinements by further theories.

A number of competing theories have been advanced to further refine our understanding of gravitational and inertial effects. The majority of theories advanced appear to seek to refine Einstein's theory of general relativity, a refinement of Newton's *law* of gravity. Only continued observations can establish which if any of them more closely approaches reality.

Ghosh's unified theory of gravity and inertia with its additional velocity component to the gravitational mass effect of matter and energy does not necessarily depend on the Einstein theory. That, in itself does not necessarily set it above any other theory. But it offers an explanation of the exact correspondence of inertial and gravitational mass without requiring any fine tuning of the universe's basic parameters like mass density and size. It offers a fine tuning of gravitational mass that affects both Newton's laws of motion and his law of gravity. And, through the latter, if not adequate by itself, offers a fine tuning of the required gravitational inputs for particular solutions to Einstein's general theory of relativity.

The Ghosh theory provides a general explanation for phenomena which otherwise require individual ad hoc explanations and predicts other phenomena not previously observed. It has not been refuted and its conclusions should be closely investigated.

* The importance of using tensors lies in their being mathematical entities that are invariant [they describe quantities in such a way that they will be measured the same by all observers regardless of the coordinate system in which they are described]. Einstein believed his theories would have been better referred to as theories of invariance rather than relativity. Minkowski was the first to show that the theory of special relativity could be cast in a vector form so that time and space were just shadows of the time-space continuum reality. And vectors are simply the lowest order tensors.

Addendum—What is Fundamental

In basic physics, at the least, a fact is not established unless it is confirmed by all known observations. In that sense, this is the fundamental of physics.

Insofar as measured observations are concerned, light seems to be the basic measurement tool. Given that, as Einstein's Theory of Special Relativity [ESR] basically concerns the measured properties of light, correspondence with ESR might well be considered the most fundamental requirement in physics.

You might remember that it is specifically based on these things:

1. The measured speed of light is the same for all observers in uniform motion with respect to each other [called inertial frames].
2. All physical phenomena behave identically the same in all inertial frames.
3. Implied at least: What is measured is all that physics is concerned with.

Edward Purcell assumed that ESR applied to electrical measurements in explaining, arguably, the most basic explanation of magnetism even when the relative speeds of the electrical charges causing the magnetism were considerably less than the speed of light. He then made his point with accepted mathematical calculations showing the magnetic force results agreed with observations.

The basic point I am trying to make here is that ESR has been shown to apply at relative speeds much less than that of the speed of light in this case, and, if so, one would think that such speeds would cause ESR effects for everything else.

Electrical inductance might be renamed *inductia* in that its effect is a perfect mathematical analog to mechanical inertia if electrical charge is substituted for inertial mass.

ESR calculations clearly show that measured inertial mass increases with the relative speed of that mass. That is, the measured inertia of the mass as determined by its relative speed. So, one could reasonably say that the cause of inertia is ESR—largely because that theory depends on the measured speed of light.

One could ask what makes measurement of the speed of light what it is, let alone why it is even finite. For all we know, the cause might be the total mass of the universe. Mach and Ghosh argue that is the cause of inertia. Ultimately, the speed of light and ESR are confirmed by exhaustive measurements which have never shown otherwise although showing any exceptions would make the reputation of anyone doing so.

Appendix
Mathematics and Individual Equations

The following again lists the four basic Maxwell equations as best I can reproduce the symbols conventionally used. I have used the word del to represent the upside down delta that conventionally symbolizes the operator del or nabla. Del is a vector so it is represented in bold text as all vectors are:

(1.) $\nabla \times \mathbf{H} = \iota \times \partial \mathbf{D}/\partial t$

(2.) $\nabla \times \mathbf{D} = - \partial \mathbf{B}/\partial t$

(3.) $\nabla \cdot \mathbf{B} = 0$

(4.) $\nabla \cdot \mathbf{D} = \rho$

Mathematically the del operator is represented in rectangular coordinates as:

$\nabla = \mathbf{i}\, \partial/\partial x + \mathbf{j}\, \partial/\partial y + \mathbf{k}\, \partial/\partial z$ where $\mathbf{i}$, $\mathbf{j}$, and $\mathbf{k}$ are the unit vectors along the x, y, and z axes respectively, **del** is a vector in form but is really only an operation until such time as the variables indicated by the partial derivatives are named.

For more specifics on del or nabla, you may visit [Link 19].

The appendix provides a detailed discussion on the meaning of the mathematics behind each of the individual Maxwell equations.

Description of the Mathematical/Physical Terms used in Maxwell's Equations

If you were to form an understanding of a customary mathematical form for Maxwell's equations, you naturally need to know the definition of each of the terms: [Vector fields are shown in bold text.]

H and **B** are both magnetic field designations where $\mathbf{B} = \mu \mathbf{H}$ and μ in this case is the so-called permeability of free space, nothing more than the degree of magnetism that free space obtains in a magnetic field, μ here a constant of proportionality between **B**

and **H**. If you wish, you may visit [Link 20] for a more general discussion of permeability.

E and **D** are both electric field designations where **D** = ε**E** in which ε in this case is the so-called permittivity of free space, nothing more than the degree of resistance that is encountered in forming an electric field in free space, ε here a constant of proportionality between **E** and **D**. If you wish you may visit [Link 21] for a more general discussion of permittivity.

The partial-derivative symbol is ∂. It is used with a function to designate a differential of a function with several variables with respect to one of those variables <u>with the others held constant</u> (as opposed to the <u>total derivative</u>, in which all variables are allowed to vary). If you wish, you may visit [Link 22] for a general discussion of partial derivatives.

I choose the term differential here to define an infinitesimal change as opposed to derivative which can be thought of as the ration of infinitesimals. If you wish, you may visit [Link 23] for a more exhaustive discussion of the term infinitesimal. I encourage you to google any term that you encounter when you are in doubt, as you can the term derivative, and Wikipedia is generally a good source in such a regard.

The isolated partial derivatives in Maxwell's equations are taken with respect to t, time in this usage. Partial derivatives with respect to space are incorporated in the vector operator del or nabla, the upside down delta, frequently shown on the left hand side of Maxwell's equations, in one equation with **H** in another with **E**. For more specifics on del or nabla, you may visit [Link 24].

The remaining term in the Maxwell equations using vector cross products is ι or iota, the current density induced in a conductor in a magnetic field. Current density is the <u>electric current</u> per unit area of cross section. It is defined as a <u>vector</u> whose magnitude is the <u>electric current</u> per cross-sectional area at a given point in space (i.e. it's a <u>vector field</u>).

If you wish, you can visit [Link 25] for more detail on current density.

The fact that current could be induced in a conductor by varying the magnetic field about it had been previously ascertained experimentally. The displacement current term, the partial derivative of **D** with respect to time is what Maxwell inferred and is said to be crucial in his prediction of then not known electromagnetic wave propagation through space—as well as his prediction of the speed of that propagation using the previously known physical constants for permeability and permittivity in free space.

The remaining Maxwell equations involving the dot products of nabla and **B** or **D** show the lack of divergence of **B** in open space and that the divergence of **D** is proportional to charge density. Charge density is a measure of *electric charge* per unit *volume* of space. If you wish, you can visit [Link 26] for more particulars on charge density.

Verbal Description of Each Equation

Given the above, definition of terms, you can interpret the meaning of each one of Maxwell's equations with the many words required as follows:

1. Most simply, a variation in a magnetic field in a space will induce an electrical current in a conductor in that space and a variation of an electric field in that space at right angles to the direction of the inducing field, that has come to be called a displacement current. The direction of the induced field would be at right angles to the inducing field in the direction that the inducing field would go if it were a screw turned in a clock-wise direction. Alternately, without regard to field direction, a given magnetic field variation in space will induce a given current density in a conductor within that field as well as a displacement current in that space. Stated, with a bit more mathematical precision: the cross product of a given variation of the **H** magnetic field in space with the vector operator nabla which forms the total derivative for that field **H** equals a given current density in a conductor within that field as well as a displacement current equal to the partial derivative of **D** with respect to time in that space.

2. Most simply, a variation in an electric field in space will induce a variation of a magnetic field at right angles to the direction of the inducing field in that space The direction of the induced field would be at right angles to the inducing field in the direction that the inducing field would go if it were a screw turned in a clock-wise direction. Alternately, without regard to field direction, a given electric field variation in space will induce a given variation of the **B** magnetic field within that space. Stated with a bit more mathematical precision: the cross product of a given variation of the **E** electric field in space with the vector operator nabla equals the negative of the partial derivative of the **B** magnetic field with respect to time within that space.

3. and 4.
I already discussed the physical significance of the remaining third and fourth Maxwell equations.

The words necessary to describe the phenomena even after the vector fields are defined are obviously many more than that necessary with the mathematical symbology. Now that we have defined the mathematical symbology, we can meaningfully exhibit the Lorentz force. That force includes the magnetic force that moving charges exert on one another: $\mathbf{F}_B = q\mathbf{v} \times \mathbf{B}$. The combination of static electric and magnetic forces on a charged object is known as the Lorentz force: $\mathbf{F} = q(\mathbf{E} + \mathbf{v} \times \mathbf{B})$. That is, the force due to the electric field is equal to the charge upon which it acts times the magnitude of the electric field and aligned with the direction of the electric field, and the force due to the magnitude field equals the charge times the charge's velocity times the magnitude of the magnetic field directed at right angles to the magnetic field in the sense described in the definition of a cross product.

Mathematical manipulation of Maxwell's equations predicts a speed of electromagnetic wave propagation in free space using the physical constants described above. That derived speed equals one divided by the square root of the product of the electrical permeability of free space times the electrical permittivity of free space [$v = 1/\sqrt{\mu\varepsilon}$]. That speed was consistent with the experimental determinations of the speed of light and suggested that light was one form of electromagnetic radiation.

References

Barnett, Lincoln *The Universe and Dr. Einstein*, with a foreword by Albert Einstein, New York, Copyrights 1948,1958

Bondi, Hermann *Relativity and Common Sense—A New Approach to Einstein*, New York, Dover Publications Inc. 1962, 1964

Chemical Rubber Company CRC *Standard Mathematical Tables* 1965

Einstein, Albert *Ideas and Opinions*, New York, Dell Publishing Co. Inc. 1973, New York, Crown Publishing Co. Copyright 1954 [There are many subsidiary copyrights for the individual papers]

Einstein, A., Lorentz, H. A., Minkowski, H., Weyl H. *The Principle of Relativity* [A collection with notes by A. Sommerfield translated by W. Perrett and G.B. Jeffery published 1923 reprinted through special arrangement with Metheun and Company LTD and A. Einstein] New York, Dover Publications Inc.,

Ghosh, Amitabha *Origin of Inertia—Extended Mach's Principle and Cosmological Consequences*, Montreal, Apeiron, 2000

Lieber, Lillian R. *The Einstein Theory of Relativity—A Trip to the Fourth Dimension*, edited and with a new foreword by David Derbes and Robert Janzen, Philadelphia Pennsylvania, Paul Dry Books, Inc, 2008 [updated from Lillian R. Lieber original 1936, 1945]

NASA has announced the results of an epic physics experiment which confirms the reality of a space-time vortex around our planet, http://science.nasa.gov/science-news/science-at-nasa/2011/04may_epic/NASA May 4, 2001

Purcell, Edward Mills *Electricity and Magnetism, Berkeley Physics Course Volume 2*, Boston, Massachusetts, et. al. McGraw Hill Inc. 1985

Sciama, D. W. *The Physical Foundations of General Relativity*, Garden City New York, Doubleday & Company, Inc. Anchor Books edition, 1969

Sciama, D. W. *The Unity of the Universe*, Garden City New York, Doubleday & Company, Inc. 1961

Skilling, Hugh Hildreth *Fundamentals of Electric Waves*, New York, John Wiley & Sons, Inc. London Chapman and Hall, LTD, 1948

Quantum Electrodynamics Sequel to
Understanding Effects Across Space

George Henry Edwards

Copyright © 2014 by George Henry Edwards
Permission is granted for anyone to quote or copy and paste any or all of the copyrighted material in this sequel of *Quantum Electrodynamics Sequel to Understanding Effects Across Space* or the URL: www.giac2002.org/qed.htm is cited as the source.

LIMIT OF LIABILITY/DISCLAIMER OF WARRANTY

The author makes no representations or warranties with respect to the accuracy or completeness of the contents of this work. The fact that a work, person, organization or Website is referred to in this work as a citation or potential source of further information does not mean that the author endorses the information the work, person, organization or Website may provide. Readers should be aware that Internet Websites listed in this work may have changed or disappeared between when this work was written and when it is read.

Cover pages, graphics and book design by Susan Edwards Gum

Library of Congress Control number: 2015901788
ISBN-13: 978-1507739952
ISBN-10: 1507739958
CreateSpace Independent Publishing Platform
North Charleston, SC

Dedicated to Marian Weiss Edwards

Introduction

This *Quantum Electrodynamics* sequel to the book *Understanding Effects Across Space* seeks to explain quantum electrodynamics, QED. Luminaries in the field have cautioned that no one understands quantum mechanics or its subset QED, so my aim is not so lofty as to do more than try to explain its findings and procedures.

QED defies our common sense that is based on everyday observations. It is not even amenable to the crisp logic that we applied to classic [pre-quantum mechanics] electromagnetism. As first exemplified by radio-activity, it is essentially characterized by randomness and the best that theory can do is predict the probabilities, the odds, of quantum events. Its predictions as to those odds however stun the imagination with their accuracy.

My explanation on the procedure followed in the theoretical predictive approach is primarily based on the non-mathematical book, *QED, The Strange Theory of Light and Matter* written by a primary founder of the approach, Richard Feynman, and another more recent book augmenting his explanation and other quantum physics theory: *The Quantum Universe*, by mathematician-physicists Brian Cox and Jeff Forshaw. The Feynman procedure is apparently that most generally followed in most applications predicting QED results.

I further discuss wave motion, in general, using the hydraulic analogy to water waves and the mathematical analogy/abstraction of complex numbers. Otherwise, other than a very top-level discussion of standard probability theory, I avoid mathematics beyond that most probably covered in your high school.

Chapter 1
Knowledge, Truth, Trust, Proof, Belief and this Book

As I've written before, I believe an individual only *knows* what others have told him or written and he believes or accepts. It is impossible for an individual to have personally made all the observations upon which to base his or her beliefs. He must evaluate whether what he is told is true or not.

To do so, he should evaluate whether the source should be trusted to tell the truth as he knows it and that the source has a reasonable wherewithal to accurately discern the truth.

He needs to evaluate the truthfulness of the source when the source even makes a statement of his actual observation not to mention whether the source is rational and not reporting such as hallucinations. He should evaluate whether the source has any rational reason for telling an untruth, that is, if the source has any conflict of interest such as gaining by telling an untruth. Perhaps even better, would the source lose in some way by telling an untruth? Have previous statements from the source been found to be true?

In the case of a mathematical or other logically based report, are there inconsistencies in the report, that is, are there any logical contradictions in the report? In the case of observational reports, are there any observations contrary to the reported observation?

Given the appropriate answers to these questions, a belief in the report or at least its acceptance is justified and may be reasonably maintained as long as contrary results are not reported or logical inconsistencies are not subsequently found.

It may be that there is not a crisp logical explanation of a phenomenon. For instance the Aristotelian premise of the *excluded middle* does not apply in quantum mechanics. Gödel has shown that there are postulates that can neither be mathematically proven or disproved. However, the odds on particular quantum results can be determined. Quantum mechanical theory predictions are frequently in *fuzzy logic* as opposed to the crisp logic—either true or false—realm. All that is *possible* to predict are the odds of an occurrence.

All that I know or accept of quantum electrodynamics or quantum theory in general, I have read in the books referenced, on the Web, or been told. I notice that, decades ago, I did take an undergraduate course in modern physics. I no longer even attempt to work out mathematical explanations.

This sequel expects no mathematical background beyond, perhaps, an exposure to that provided in high school or touched upon in this book's prequel *Understanding Effects Across Space*, hereinafter referred to simply as *Understanding*. Except I do provide some very basic elements of three areas of mathematics perhaps not generally taught in high school—vectors, complex numbers and probability. In doing so, I hope to shorten the words needed in their special application to Feynman's explanation of his approach to theoretical quantum electrodynamics procedures, *QED*.

This book is generally my distillation and interpretation of the content of the books in the references—in conjunction with further references to World Wide Web sources that appear to be consistent with the contents of these books. This book sequel is an attempt to explain QED concepts, theoretical approaches and salient interpretations with minimal mathematics.

The terms electrodynamics and electromagnetism are nearly synonymous in that both are concerned with the physics of electricity and magnetism.

The word light, as used in this sequel, generally refers to all electromagnetic radiation including the everyday phenomena of light, radio and x-rays.

Any errors are, of course, mine. Please e-mail me at edwardsgeor@gmail.com with any corrections that you believe should be made.

There are many sources available going into mathematical detail for those who want to try to enhance their understanding with them—for instance, starting with standard textbooks on modern physics. And mathematics is essential for predicting detailed real world application results.

Math provides a shortcut for thinking by using theorems or results long established to have no internal contradictions. The results of mathematical models and their correspondence with real observations may suggest thoughts for other investigations and model enhancements. The practical usefulness of mathematics in the real world lies in its repeated validation providing engineers and other practitioners reliable tools for predicting reliable results in specific applications.

On the other hand, for many, even those proficient in mathematical manipulation, studying mathematical details can obscure physical understanding. Use of mathematics can be similar to the use of digital computers in that it can most efficiently obtain accurate predictions from established mathematical models without the user seeing or even being aware of, the detailed considerations or manipulations involved. This excerpt attempts to, at least, give a basic understanding of QED concepts and procedures.

Chapter 2
Distillation of Quantum Electrodynamics [QED]

Quantum electrodynamics is the quantum view of electromagnetism as opposed to the so-called classic [pre-quantum theory] view in which physical phenomena are considered continuous. It is a subset of quantum mechanics in which all physical phenomena are considered to be made up of separate irreducible quantum particles, quanta.

Experiments indicate that quantum results are basically and irreducibly random so that Aristotle's crisp logic of the excluded middle does not apply. Therefore the usual requirement for a theory to be mathematically and so crisp-logic-consistent does not necessarily apply in the quantum realm. The physical criterion of assured predictability is all that remains.

In the later 19th century, so-called *radioactivity* was observed in which certain substances emitted particles and rays in such a *random* fashion that only knowing the half-lives of the radioactive substances was possible, and cathode rays were found to be streams of electrons.

In the early 20th century, light, that is any form of electromagnetic energy, was found to be quantized in particles called photons that could constructively and destructively interfere with themselves like waves. Postulations that all forms of matter had wavelike as well as particle characteristics were experimentally verified. An absolute limit on the accuracy of determining certain joint quantities such as position with momentum in quantum mechanics was established. This meant that the best that any quantum theory, based on this absolute indeterminacy could do was determine the odds of a particular quantum event occurring.

The first overall QED theory advanced was in the form of the non-relativistic Schrödinger wave equation. It was almost immediately re-interpreted so that quantities that we now call *probability amplitudes* were to be squared to give probabilities. It was later refined to include the effects of special relativity.

Current theories dispense with the idea of the **fields** of classic electromagnetism other than as entities caused by the conglomeration of electrons such as in antennas or transformers; they propose that such phenomena are merely the results of individual electrons interacting over space with other electrons via emitted and absorbed photons

The current apparently universally used, and immensely successful, procedure for predicting quantum electrodynamic results combines the individual probabilities for **all** possible paths in time and space for quanta—each having the potential for wave-like interference or re-enforcement. These with an *orgy of quantum interferences* produce an overall probability for an event. The possible paths include all those zigging and zagging in quanta directions in addition to those of averaged conventional light waves and at speeds greater than or less than that conventionally ascribed to light waves. The *orgy of interferences* results in the conventional results having their overwhelming probability.

It turns out that the considerations for electron hopping to any place in time or space are very similar to those for photon hopping except a factor related to the electron mass and velocity must be included. Interactions between electrons and photons, be they either emission or absorption, can be handled identically in an elementary fashion, according to the essentially non-mathematical book *QED*, written by a primary author of the *sum over histories* [of paths] approach, Richard Feynman.

"The theory of quantum electrodynamics is also the prototype for new theories that attempt to explain nuclear phenomena."

In QCD [quantum chromodynamics, a predictive approach similar to QED is used except other *quanta* than electrons and photons are considered. QCD and QED are primary ingredients of the so-called *standard theory* of physics that covers the rest of the present-day theory of all of physics except for gravity.

The essentially non-mathematical book, reference that includes QCD to any real extent as used in this sequel, *The Quantum Universe*, in "References," goes into a very top-level discussion of how the concept of physical action is used for that theory. That includes the mass of the particular particle involved. As the book *QED* considers only electron quanta, only an electron's mass is involved and considerations for other mass calculations are not needed.

A primary question that remains is the interpretation of what the predictions mean in the real world. For instance, are they simply extremely accurate mathematical computation devices or are *probability* waves real physical entities?

Chapter 3
The Beginnings

1869—Johann Hittorf observed what were subsequently called *cathode rays*.

1896—Henri Becquerel discovered what is called *radioactivity* in which rays or particles are spontaneously generated randomly although the odds are constant over time. That is, it has proven impossible to predict when a particle or ray would be generated but the rate of generation has been discovered to be constant over time.

1897—J. J. Thompson showed cathode rays were composed of negatively charged particles that were subsequently called *electrons*.

1901—Max Planck precisely predicted the radiated energy from a *black body* over all frequencies by postulating it to be the combination of radiation energies from individual oscillators each with a discrete energy equal to its frequency, f, times a constant, h, that became called *Planck's constant* or the *quantum of action*. That is, each energy contribution E was equal to h times f where h is Planck's constant and f is the frequency in cycles per second, currently called *hertz*. Neither he nor anyone else has been able to describe observed results completely accurately assuming any kind of continuous energy distribution.

1905—Albert Einstein postulated that photo electricity in which electrons were emitted from a *photoelectric* material only when the frequency of the light striking the material exceeded a certain value was due to the energy of individual light quanta such as the energy quanta postulated by Planck. He wrote: "According to the assumption to be contemplated here, when a light ray is spreading from a point, the energy is not distributed continuously over ever-increasing spaces, but consists of a finite number of 'energy quanta' that are localized in points in space, move without dividing, and can be absorbed or generated only as a whole." Of course, the light quanta are now called *photons*.

Planck was not impressed with his constant as he considered it merely a *fudge factor* that was needed to accurately *fit* observation. It's worthwhile to carefully consider what his constant said—that rather than radiated energy being continuous, radiations are discrete quantities that have energies directly proportional to their frequency. Planck's constant is the constant of proportionality that allowed that proportion to be cast as an equality

Einstein's words encapsulate quantum theory and quantum radiation insofar as light is concerned except for its randomness. The diminution of the radiation proportional to the inverse square of the distance from its source is an accepted reality in both classical and quantum theories. Einstein never accepted the randomness associated with the currently accepted quantum theories. He famously opined that "[God] is not playing with dice." His application of the photon electron interaction in photo-electricity anticipated interactions in such as Feynman's theory.

Thomas Young, James Clerk Maxwell and Heinrich Hertz had proven that light was composed of waves despite Newton's idea that light was composed of particles that he called *corpuscles*. So, the postulation that light was composed of light quanta rather than being the result of strictly a wave characteristic was initially resisted, but further observations established that light had both particle and wave characteristics.

In 1927, Linton Davisson and Lester Germer reported their conduct of a two slit experiment with electrons that confirmed de Broglie's 1924 postulation that electrons have wave-like characteristics.

Chapter 4
Waves and Probabilities

For those of you for which this is all *old hat*, feel free to skip this chapter.

Probability Principles Adequate for Quantum Physics Explanations

The probability of an event happening is the number of ways it can happen [the number of so-called *successes*] over all the possible ways it could happen. So the probability for a successful event that is absolutely certain is 1 and the probability of no *successful* event ever happening is 0.

If r_1 signifies the successful results in one case and r_2 signifies the successful results in another case, and n is the number of all possible results, then the probability of success in one case is r_1/n and the probability of success in the other is r_2/n.

The probability of one or another of the successes happening is the sum of the probabilities of the individual successes happening.
For example, $(r_1 + r_2)/n = r_1/n + r_2/n$.

On the other hand, the probability of successes happening together is the product of the probabilities of the individual successes happening. For example, the product of r_1/n times r_2/n is the probability of two successes happening together.

In the case of two dice for instance, the probability of getting a particular number of spots on either of two *fair* dice, [that is dice in which the probability of getting any particular number on a single roll of either die with 6 spots is 1/6], the probability of getting any particular combination of total spots on the two dice—either when thrown together or when thrown in sequence—is 1/6 times 1/6 or 1/36.

To then determine the probability of getting a *particular* total number of spots by adding the dots on the two dice, one has to identify the number of instances in which the sums will yield that particular total number of spots, and divide by the 36 possible two-dice combinations to find the probability:

For a two—1,1   one way   probability 1/36

For a three—1,2 | 2,1   two ways   probability 2/36

For a four—2,2|3,1|1,3   three ways   probability 3/36

For a five—4.1|1,4|3.2|2.3   four ways   probability 4/36

For a six—5,1|1,5|3,2|2,3|4,2 five ways   probability 5/36

For a seven—5,2|2,5|4,3|3,4|6,1|1,6   six ways   probability 6/36

For an eight—5,3| 6,2|4,4| 3,5|2.6   five ways probability 5/36

For a nine—5.4|6.3|3.6|4.5   four ways   probability 4/36

For a ten— 5,5|6,4|4.6   three ways   probability 3/36

For an eleven—5,6|6,5   two ways   probability 2/36

For a twelve—6,6   one way   probability 1/36

Complex Numbers

The general representation of a complex number is the sum of a real number with a so-called *imaginary* number. An imaginary number is the product of a real number and i where i symbolizes the square root of negative one; it is called an *imaginary number* because negative square roots have no *real number* values.

The name *imaginary number* may be unfortunate. It has no mystic connotation. With the previously established, so-called *real* numbers it extends the original numbering system, that could be used to denote all the points on an infinitely long line, to the complex number system which can be used to denote all the points on an infinite plane. [Similarly, there is nothing insane about so-called *irrational* numbers that are only the subset of real numbers that can not be expressed as *ratios*.]

In a rectangular coordinate system, the real part of a complex number is conventionally represented as a projection on the horizontal axis and the imaginary part on the vertical axis. The overall imaginary number can be represented by the pair of numbers associated with its projections on the coordinate axis in which every point on the complex plane is then represented as the intersection of perpendicular lines from the horizontal and vertical axis projections. Alternately, the overall imaginary number can

Waves and Probabilities   133

be represented by a directed arrow with the component projections on the coordinate axes—that is a vector. In any case, complex numbers abide by all the rules of the algebra of real numbers except for order.

The sum of any two vectors is a vector formed by the sum of their individual horizontal and vertical components or projections on the coordinate axes and has an absolute value that accords with the Pythagorean theorem, that is, it is equal to the square root of the sum of squares of the absolute values of the horizontal and vertical components.

Because the product of i times i equals -1 by definition [$\sqrt{-1}$ times $\sqrt{-1}$ = -1], the multiplication of two complex number vectors, both at 45 degrees with respect to the horizontal coordinate axis will result in a complex number vector at 90 degrees with respect to the horizontal axis. Any complex number vector with equal real and imaginary components would be a 45 degree angle vector example. For instance, one could form two such vectors and form their product by choosing one factor to be 4 + 4i and the other to be 3 + 3i. Their product can be computed as follows:

$$
\begin{array}{r}
4 + 4i \\
\times\,(3 + 3i) \\
\hline
12i - 12 \\
12i + 12 \\
\hline
\end{array}
$$

24i, that is 24 units at 90 degrees with the horizontal axis; that is, the angle is the sum of the individual 45 degree angles that the two original vectors had with respect to the horizontal axis.

It turns out that the product of *any* two complex number vectors is a vector with a magnitude equal to the product of their real parts and with a direction equal to the sum of the angles that the vector factors have with the horizontal axis.

The arrows that Richard Feynman and Brian Cox and Jeff Forshaw refer to in their books [see "References"] are just such vectors that are added or multiplied using the same rules that the long-standing probabilities described above have used.

The results however are what are called *probability amplitudes* that in quantum theory usage have to be squared to find actual probabilities.

In his book, *QED*, Feynman writes that the instances where you would multiply probabilities you multiply probability amplitudes. That is, in cases in which you are dealing with compound events that can be broken down into a series of steps you multiply; in cases, where events happen independently, you just add.

Wave Representations

As in *Understanding*, where hydraulic analogies were extensively provided, water waves are analogous to electromagnetic waves or de Broglie matter waves. Spinning mathematical vectors or *arrows* that can be conveniently represented by complex numbers are in turn analogous abstractions for any wave. They would be represented as turning completely around in one complete period with a frequency corresponding to that of the wave, moving at the same velocity as that for a wave in a media. Electronic wave quanta, photons, could be represented as moving at the speed of light and diminishing in amplitude inversely as the square of the distance from a source.

Such represent single frequency, that is, with light, single color, or monochromatic radiations. Real electromagnetic waves may not spin in one plane only, that is they may be other than linearly polarized—confined to a given plane along the direction of propagation, but polarization, is most frequently ignored for convenience in dealing with arrow analogs including what are called electric or magnetic field vectors.

Indefinitely continuously turning vectors represent indefinitely continuous wave propagation.

In the real world, there may be interactions with any wave that may interrupt the wave or even cause it to rebound in such a way as to cause a counter wave or eddy current in an opposite direction and phase. Water waves in an otherwise steady stream may be interrupted by a boulder or by *breaks* in an ocean flowing in almost random directions when they approach a shore. A so-called *wavicle* of limited duration, representing a particle, is made up by a combination of waves of many frequencies; furthermore its fundamental frequency changes with its energy divided by the planck constant, h. For some reason, Feynman in his QED book does not mention this at all, instead simply invoking an imaginary stop watch with no physical rationale.

Water and Electromagnetic Waves

A water wave on a surface is evidenced by a pile-up of water in one location that, due to the force of gravity, provides a downward pressure that results in a horizontal transverse pressure that in turn will result in a horizontally and time displaced vertical transverse pressure. The pressures result in traveling waves of vertical and horizontal water displacement.

Similarly an electromagnetic wave is evidenced by an excess of an electric field in one location, that causes a displacement current and hence a transverse magnetic field in space in a perpendicular plane to that location. That in turn will result in a time displaced electric field in a plane perpendicular to that plane [the same plane as the previously occurring electric field]. The varying electric and magnetic fields result in perpendicular field displacements at any specific location.

Both cases represent examples of action and reaction in accordance with Newton's law that for every action there is an equal and opposite reaction [but not necessarily in the same location in either time or space].

The same wave principles apply to all waves including the de Broglie waves generated by the momentum of any physical particle in proportion to its momentum. The deBroglie wavelength, $\lambda = h/\rho = h/mv\sqrt{(1 - v^2/c^2)}$ when taking into account the effect of special relativity.

The foregoing alone may adequately describe to many of you almost everything about waves and their vector representation. But let me write a bit more about waves for those of you who may not have previously been closely involved with their descriptions. When speaking of waves, *period* is the amount of time, typically measured in seconds, that it takes for one complete cycle of a wave; *frequency* refers to the number of recurring cycles in a second. The sine wave is, perhaps, the most common curve representing smooth repetitive motion used in mathematics, physics and engineering. The following figures illustrate wave length when used in representing sinusoidal wave propagation over space and sinusoidal period with respect to frequency over time. The absolute *amplitude* for the waves shown in these figures is two units. The term for frequency in electricity used to be simply called the number of cycles per second or cps [Charles P. Steinmetz : >)]. Nowadays that term of measurement is called *hertz*.

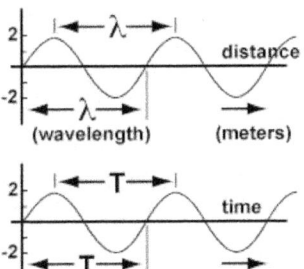

We know from classical physics that the amplitude of electromagnetic waves in free space varies inversely as the square of the distance from the source; this seems reasonable if one considers that the surface of a sphere into which a wave expands varies inversely as the square of the distance from its center—most importantly this accords with measurements made from an omni-directional radio antenna or other electromagnetic wave projector, including a light, source.

Chapter 5
QED Theories

Equations were developed that had varying degrees of success in predicting the results of quantum interactions. The currently generally accepted Feynman procedure yields amazingly precise predictive results.

In 1924, Louis de Broglie postulated that electrons, not just photons have wave characteristics. He further extended this idea to postulate that all particles have associated waves.

In 1925, Werner Heisenberg published the [quantum] Uncertainty [indeterminacy] Principle.

In 1925, Erwin Schrödinger formulated his non-relativistic wave function equation [published in 1926] that he said merely formalized the Einstein-de Broglie quantum postulates. Werner Heisenberg, Max Born and Pascual Jordan introduced a matrix formulation of quantum mechanics. Heisenberg published the [quantum] Uncertainty [indeterminacy] Principle.

In 1926, Schrödinger proved the mathematical equivalence of his wave mechanics and Heisenberg's matrix mechanics. The Cox and Forshaw referenced book says that Dirac eventually proved "in all cases [they were] entirely equivalent." Max Born observed that squaring the probability density function [now called, the probability amplitude] in the Schrödinger equation would yield the actual probability. That interpretation was initially opposed by Schrödinger and Einstein, but Born won the 1954 Nobel Prize for it and it's now generally accepted.

In 1927, De Broglie postulated a causal pilot wave interpretation of quantum mechanics that was set aside by the physics community for the stochastic, so-called *standard model*, *Copenhagen* interpretation. However the pilot wave approach was independently re-introduced and refined by David Bohm in the 1950's.

In 1928, Paul Dirac refined the Schrödinger equation so that it included the effects of special relativity.

In 1935, Albert Einstein, Boris Podolsky and Nathan Rosen wrote the *EPR* paper that was meant to show that the Copenhagen interpretation of quantum mechanics is incomplete. Many tests continue to appear to uphold the Copenhagen interpretation in this regard so it is commonly known as the *EPR paradox* and laid to what is called entanglement.

In 1940's, the formulation of precise quantum electrodynamic calculations was made by R.P. Feynman, F. Dyson, J. Schwinger, and S.I. Tomonaga.

In 1948, the complete path integral formulation was developed.

In 1952, David Bohm independently advanced Bohmian mechanics, which is currently also called the de Broglie-Bohm theory, the pilot-wave model, and the causal interpretation of quantum mechanics. The theory is non-local so the *EPR paradox* is not paradoxical in this theory. It is a version of quantum theory originally advanced by Louis de Broglie in 1927.

In 1965, Richard Feynman, Julian Schwinger and Sin-Itro Tomonaga received the Nobel prize for their work developing the QED theory.

Chapter 6
Feynman Path Integral Approach

In 1983-5, Richard Feynman published *QED*, a basically non-mathematical book that is the primary source for the material written on his quantum electrodynamics *procedure* here. Feynman was also a renown teacher. I highly encourage reading his book that is replete with graphic illustrations. I am reporting it from the point of view of one with much less mathematical background and competence than he had and so might further the understanding of others who are not already exceptionally well-versed in these areas.

He explained his procedure with this statement as to its accuracy: "If you were to the distance from Los Angeles to New York to this accuracy, it would be exact to the thickness of a human hair."

You may choose to click [Link 1] *QED for beginners* from Stanford University] that, although also abstracted from Feynman's book, provides more detail than this sequel.

[When using the printed book, if you have access to a computer, and wish to read augmenting material associated with one of the numbered links in this sequel, visit giac2002.org/qed_links.html. And once there, click the word "here" in the associated link number "Click here" phrase.]

Feynman's book as well as the book by Brian Cox and Jeff Forshaw: *The Quantum Universe* use the terms *little arrows* and *big arrows* when referring to complex number vectors and their arithmetic in forming probability amplitudes. If you are not yet aware of the very basic concepts of these represented here as well as the very basic mathematical probability concepts involved, read the Chapter 4 sub-section "Elementary probability principles adequate for our quantum physics explanations" in this sequel. Or you can read Feynman's book *QED* which lays all the vector/probabilistic ideas out in great detail with illustrations using *little arrow* and *big arrow* explanations. The *little arrows* refer to the individual probability amplitude vectors; the *big* or *final* arrows to the accumulation of all the *little arrow* vectors over all time and all space.

140   Chapter Six

In his book, Feynman represents the *probability amplitudes* for individual photons by what are generally called *complex number vectors* but he calls *little arrows* whose direction rotate once per cycle of the particular frequency that they simulate as described in our sequel Chapter 4 sub-section "wave representations." Feynman gives what turns out to be the elements of his overall approach, by showing how his methods can be used to explain, the fact that the *angle of reflection* of a light beam from a mirror equals its *angle of incidence* to the mirror.

He proposes a thought experiment in which a photon at a particular frequency and a photomultiplier detector are placed in front of the face of a mirror. Remembering that the little arrows rotate with time as they propagate, their length when they reach the mirror varies with the distance from the source to the mirror, the vector sum of the *little arrows* along all the *possible* paths the photon can take between the source and the detector will maximize when the total path taken is least (also the least time path). That is, when the resultant combined *final arrow* or *big arrow* arrived at by finding the vector sum of all the arrows [including those interfering to some degree with each other (their vector differences)] is the greatest. In the case when the angle of incidence is equal to the angle of reflection. His book illustrates this with the following graphic where S indicates the position of the source, P the position of the photomultiplier detector and Q just a screen to block the paths between the source and detector:

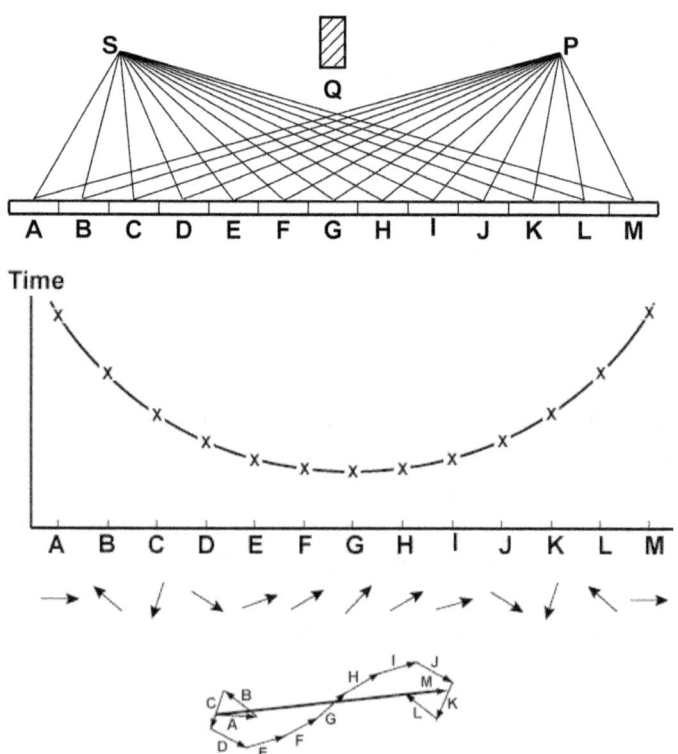

Feynman Path Integral Approach   141

The graphic helps make it clearer that vectors corresponding to the photons that are near the equal angle of incidence/refraction are in greater commonly directed alignment than those further away from that angle which more often partially or totally subtract/interfere with each other. Overall, then, the effects of a shower or spray of photons reflecting from a mirror will positively combine near the angle of incidence and reflection and partially or totally interfere with each other at angles further displaced from that angle. In his words: "It is evident that the major contributors to the final arrow's length is made up by arrows E through I whose directions are nearly the same because the timing of their paths is nearly the same."

Using similar elementary examples Feynman's book shows how quantum representations can be used to explain how a diffraction grating, refraction, mirages and lenses work. Any of these effects can be explained by classic electromagnetic theory although perhaps not as easily—especially by using the approach that light travels by the shortest path, the path that requires the least time.

Both books use the position of a postulated rotating stopwatch or clock hand without, at least initial, overt explanation as an analogy or abstraction for the individual quantum probability amplitudes which they frequently call *little arrows* and the accumulated final probability amplitude for the overall event a *final arrow* or maybe a *big arrow*.

As one would anticipate, the rotation rate of the little arrows when dealing with photons of a given color is the same as the classical frequency of light for that color. The rotation rate for free electrons depends on their velocity and mass; the mass of all electrons are the same, but that mass coupled with an electron's velocity is determined by a formula not provided in Feynman's book. He uses the term n.

You may choose to click [Link 2] to view a Web article: *Quantum Stopwatch for the Free Electron*, by Edwin F. Taylor—a further discussion on determining the theoretical imaginary stop watch rates for free electrons.

[Again, when using the printed book, if you have access to a computer, and wish to read augmenting material associated with one of the numbered links in this sequel, visit giac2002.org/qed_links.html. And once there, click the word "here" in the associated link number "Click here" phrase.]

I have pointed out that the rotating stop watch or clock hands are analogies/abstractions of waves. When they align in the same direction they add in the same fashion that waves with aligning crests reinforce each other, when they align in opposite directions they subtract in the same fashion that simultaneous crests and troughs of waves interfere with each other. I show how the vectors, as waves would, will combine to form the square root of the sum of the squares of their individual rectangular coordinate components.

Ultimately, the books mention that the *little arrows* and the *final arrow* are really the absolute values of complex number vectors. In the *firmly established standard* mathematical theory of probability, one adds probabilities if the resultant probability is any of the individual probabilities while one multiplies probabilities if the resultant probability is a combination of all the individual probabilities.

In QED, one has to combine the little arrows, the probability amplitudes, in *every way* that an event can happen in the same way that probabilities combine, that is, adding them if they can happen in alternate ways and multiplying them if they represent, quoting Feynman: "compound events—events that can be broken down into a series of steps, or events that consist of a number of things happening independently." The composite combination of *little arrows* or complex number vectors will result in the final probability amplitude [*big arrow* or *final arrow*]. Note the complex number vector multiplication includes addition of the *little arrow* vector angles so the product of the probability amplitudes which are less than one results in shrinks and turns. Again, probabilities are the squares of the derived probability amplitudes.

Feynman explains his concept of *sum over histories* method [also *called the path integral method* or *the sum-over-paths method*]. It considers all the paths that quanta could possibly take]. This method forms the basis of his detailed theory.

The basic idea is to first establish absolutely ALL the detailed possibilities for a quantum event including every possible path a quantum can possibly take [including zig-zags or arcs of any form] in time and in space and then sum them in the algebraic sense—that is, the sum of oppositely directed quantities may also include their vector arithmetic differences. If only a finite number of paths are considered, the prediction may fail to be precise, but the consideration of a further number of paths will yield a more precise prediction. QED theory, fully utilized, is meant to consider the entire infinite number of possibilities and so precisely predict an absolutely accurate prediction of the odds of a complete event.

This procedure accounts for the fact that quantum waves, as any other waves, can interfere with each other as well as re-enforce each other. Feynman does not go into the method for efficiently computing the infinite sums required for absolute precision.

One might wonder how his theory handles the paradoxical two slit experiment in which it has been experimentally shown that a **single** photon [or other quantum particle] will produce an *interference pattern* if suitably displaced slits are placed between the photon source and a detector—even though the detector always shows individual photons reaching the detector one at a time).

You may choose to click [Link 3] to view a demonstration that shows the *double-slit experiment* results using a slightly different instrumentation than specific double slits.

[Reminder: with the printed book, visit giac2002.org/qed_links.htm and click the word "here" in the associated link number "Click here" phrase.]

Once one detects which slit the photon went through, the interference pattern disappears.

To use the Feynman procedure correctly—surprise—one must model the presence of a detector to determine which slit the photon went through to define *the complete event* clearly. Then the *model* will show a 50% probability for the photon to be detected in either detector—even though it will in reality be detected in one detector or another—100% probability.

In his book, Feynman wrote: "I have pointed out these things because the more you see how strangely Nature behaves, the harder it is to make a model that explains how the simplest phenomena actually work."

Feynman does not limit the possibilities to be considered to the speed of light, light traveling in straight lines or particles moving forward or backward in time. It turns out that his methodology through what another author calls an *orgy of quantum interferences* causes the probabilities to converge to the conventional ones observed.

One might wonder how a quantum particle can *sniff* or *smell*, as Feynman characterizes it, the most probable path to follow. One would think that Nature is not going through the painstaking calculations that we need to use. Feynman and others indicate this does not differ from the *action* principle in classic physics in which an object needs to somehow *decide* the path to take to minimize the difference between potential and real [such as kinetic, thermal, etc.] energy.

As the Feynman approach, includes the consideration of the future possibilities of all complete paths, an outsider might wonder if these have to be [in Feynman's words] "*sniffed*" or a "photon *making up its mind.*"

Could a pilot wave [or whatever] more simply merely *sniff* the very next upcoming instant to determine whether it would yield a closer approximation to equality of the potential and actual energy with the current instant and in what direction and guide the quantum accordingly, an instant at a time? Or maybe a quantum moves an infinitesimal amount in the direction of a possible path and determines whether that particular path provides a decrease in the difference between the potential and *real* energy or not and in an infinitesimal instant or instants *corrects* to achieve the necessary equilibrium. Or simpler yet, merely achieve its full potential in the form of *real* energy?

Who knows how Nature [to personify it] proceeds? It doesn't seem likely that it follows the Feynman *sum over histories* approach of an infinite number of possible paths. That approach may be just the best that our brightest minds can come up with to calculate and predict results.

Feynman's 1965 lecture to the Nobel committee when accepting his prize "The Development of the Space-Time View of Quantum Electrodynamics" shows the gargantuan efforts even he had to expend: "I worked on this problem about eight years until the final publication in 1947." His talk reveals that even the most profound advances most frequently result from trial and error. You may choose to click [Link 4] to read his Nobel address in its entirety.

[With the printed book, visit giac2002.org/qed_links.htm and click the word "here" in the associated link number "Click here" phrase.]

Planck had kept trying different approaches until he came up with the quantum idea that accurately predicts experimentally determined heat radiation versus frequency. Historically methods for predicting phenomena were often just finding what works, like the angle of reflection is equal to the angle of incidence before it was realized that light takes the path of least time or Feynman's photon explanation.

The physics community has learned that any theory may eventually be further refined. Bitter experience has cautioned it not to call any theory a *law*, regardless of how well it is buttressed with further observations and experiments.

There are many attempted interpretations of quantum theory. if you would like to see a catalog of many instances, click [Link 5]. These instances include the so-called *many worlds* interpretation in which every possibility is actually realized in alternate universes.

[With the printed book, visit giac2002.org/qed_links.html and click the word "here" in the associated link number "Click here" phrase.] .

Arguably the least controversial of these interpretations, other than the Copenhagen or *standard theory*, [that states "that the nature of the underlying reality is unknowable and beyond the limits of scientific enquiry"] is the Bohm or the de Broglie-Bohm, *pilot* wave, interpretation. Apparently, this last is identical as to predictions as the Feyman or the *standard theory*. Its most obvious difference is that the Bohm theory asserts that the wave and particle aspects are both real—the particle goes a distinct way as does the wave. The particle, as if it included a receiver that can pick up guidance information, and is able to control the particle based on the remote information to cause the particle to follow the *sniffed* path.

In *QED*, Feynman wrote that the quantum *smells* the proper path of least action and the photon *makes up its mind*. Explanations of the *standard theory* say the choice of the quantum path may be thought of as trying every path simultaneously. The book *The Quantum Universe* [see "References"] says: "we are proposing that the electron wave spreads out to fill the universe in an instant."

Yet there are those who have no use for the Bohm causal interpretation because it assumes instantaneous action or reaction even though the Aspect results and other EPR tests at least suggest instantaneous *entanglement* in that once some properties including polarization are determined locally, an *entangled* particle, even one beyond that that can be reached at the speed of light, responds as it must due to the entanglement.

The interpretation that seems most reasonable to me, only an engineer, not a professional physicist, is that QED theory is a mathematical model to determine the odds of a particular event, nothing more. In the paradoxical two slit interference experiment, the model when properly set up gives odds only. Once the fact that a quantum has gone through a particular slit, that is a certainty—even though the a priori odds as determined by the properly set up theory have not changed. As discussed earlier, when the particular slit, the quantum goes through is not established, in theory or as measured, the detected quantum exhibits an interference pattern.

From this point of view, there is nothing mystical about the Feynman process being able to make its probability predictions instantaneously over all space and all time. A quantum is neither more than one place at once or over the entire universe simultaneously, only its possible probabilities are. If Nature picks a quantum path by sampling all paths instantaneously or by, such as say, exploring paths for an infinitesimal instant to determine whether the integral of the difference between the potential and real energy for that instant, the *action*, is minimized and in which direction. From a predictive calculation point of view, what difference does it make?

Chapter 7
Synopsis: The book QED, The Strange Theory of Light and Matter

The Feynman book QED is *right from the horse's mouth* of a QED procedure developer himself.

His book sets up his overall *sum over paths method* by showing how the accumulation of all the light quanta, *photons*, reflecting off all parts of the surface of a mirror will combine to maximize when the angle of reflection is equal to the angle of incidence. This equality has been known since antiquity [according to Wikipedia, established by Hero of Alexandria (AD c. 10-70)] and mathematically explained by conventional wave theory. QED theorists and physicists overall, postulate that individual quanta have wave-like characteristics.

Feynman's method postulates that both photons and electrons hop from place to place throughout all time and space in a random manner. It evaluates the possibility of going faster or slower than the conventional speed of light or of following the conventional straight path of light although the *orgy of interferences* of the quanta amplitudes reduces the probability of any such occurrences to near zero except for very small departures from the conventional values. His method determines the probability of any particular hop allowing the possibility of any zig, zag or looping back on itself including backwards in time.

Observations of an electron moving backward in time would appear the same as a positron—having the identical positive charge as that of an electron moving forward in time. Positrons have been observed for very brief intervals before they combine with an electron and mutually annihilate, to form a photon. "Every particle in Nature has an amplitude to move backwards in time, and therefore has an anti-particle."

His *QED* book states that disregarding polarization and taking into account the possible departures from the conventional speed of light when the difference in time between observations does not exactly equal the difference in spatial location this procedure, applied to photons "explains all about optics; that's the entire theory of light." The theory applies exactly the same for electrons hopping from place to place except that a number derived by a formula not derived or explained in his book that he calls *n* must be squared and multiplied times the probability that would be determined for electrons. This "enables all out calculations to agree with experiment."

The quantity *n*, applicable to electrons, could be likened to the proportionality constant *m* required in classical [pre-quantum] physics to make such as Newton's law of gravity [or Einstein's theory of general relativity] to yield accurate predictions for measured results. The QED technique is applicable to other quanta than electrons with different factors than *n*. The reference by Cox and Forshaw discusses this in more detail with a top-level discussion on the physical principle of *action* and associated mathematics. The book *QED* and this sequel are primarily directed to electrons and photons outside atomic nuclei so this is not discussed, other than peripherally, in that book or here.

The discussion of these constants only describes some of the details in application of the theory. Other than that, they offer nothing substantial to understanding the description of Feynman's technique.

However, the book *QED* tells of one other separately formulated number that Feynman calls a *junction* or a *coupling* that is used to calculate an action of either an electron's emission or absorption of a photon—"it's just a number! This junction number I will call *j*—its value is about -.01; a shrink to about one-tenth and half a turn." It is used as a multiplier whenever a photon is either emitted or absorbed. *j* is sometimes called the *charge* of a particle that is chosen to accurately predict the observation in this usage; it is not the conventionally measured electrical charge.

The *QED* book introduces the so-called *Feynman* diagrams that are almost universally used by physicists as computational aids: "[E]very straight line gets an E(A to B), every wavy line gets a P(A to B) and every coupling gets a *j*. Thus [in a particular book graphic example], there are six E(A to B)'s, two P(A to B)'s and four *j*s—for *every possible* 5, 6, 7 and 8! That makes billions of tiny arrows that have to be multiplied and then added together."

"[T]he rules are simple, but you use them over and over. So our difficulty in calculating comes from having to pile so many arrows together. That's why it takes four years of graduate work for the students to learn how to do this efficiently. . . . (When the problems get too difficult, we just put them on a computer!)"

Chapter 8
Synopsis: The book The Quantum Universe

The referenced book, The *Quantum Universe*, by mathematician-physicists Brian Cox and Jeff Forshaw as its title suggests touches upon overall quantum mechanics predictive procedures in addition to those essentially explained by Feynman's book *QED, The Strange Theory of Light and Matter* which despite the title use of the term *matter* concentrates on electrons that, although constituents of matter, do not encompass other quanta than electrons and photons. I will give you a glimpse of what seems to me to be most directly pertinent for our purposes, but, if you have additional interest, this book is well worth your independent reading along with Feynman's book. It covers more things in much greater detail than I do here.

The Quantum Universe continues with Feynman's attempt to try to explain the concepts behind the underlying abstract mathematical equations that otherwise would take years to learn. It proceeds in the same basic manner as Feynman. *The Quantum Universe* uses imaginary clocks and their hands rather than stop watches without really explaining their physical basis. By using these analogies both books avoid having to explicitly explain and so concern a reader about what the revolution or winding of the hands refer to physically. I touch on this a bit in this sequel.

Both books are concerned with the **probabilities of events** only. Therefore in my opinion, any statements by me or others which speak of the probabilities as actual quanta are wrong. For instance, an electron is neither *at two places at once* nor at *all places* nor *any or all places in the universe* at once. Quantum theory is used only to calculate the probability of a quantum possibly being there. There is nothing mystical about it although the possibilities investigated or the probabilities may not accord with our common sense based on our everyday experiences.

The book *The Quantum Universe* is often written as if a quantum particle is the probability itself. "[W]e have to draw an infinite number of clocks, one at every conceivable point in space. . . Allowing the particle to be anywhere at all is equivalent to assuming nothing about the motion of the particle. . . . Our particle will simultaneously be both a nanometer away and also a billion light years away in the heart of a star in a distant galaxy. . . [W]e are proposing that the electron wave spreads out to fill the

Universe in an instant." However, in the context of the book as a whole the authors use of the word *particle* as opposed to *probability* is meant only to be a figure of speech.

Both books tend to just give a set of rules using those analogies without a preponderance of explanation. This is similar to how abstract mathematics are frequently used by physicists, By trade; physicists are not generally concerned with the proofs of the mathematical theorems they use. Unfortunately, the *follow these rules approach*, although necessary, is not as intellectually satisfying, to me at least, as one might hope. This sequel offers no better approach but provides a little further explanation of the physical roots of some of these rules and more concisely discusses the approaches taken by the books.

Where Feynman's book gave an off-line derived formula and numbers applicable to electrons, the quantum universe book, in addition to other things, expands its explanation to quanta with masses other than those of electrons. You may stop reading this if you wish, as the following is not essential to a strictly QED explanation. But it may be of interest and give a bit more insight.

The quantum universe book introduces the physical concept of *action* that was actually central to Feynman's mathematical procedure. The *action* in quantum mechanics corresponds to "a particle hopping from one place to another as the mass of the particle multiplied by the distance of the hop squared divided by the time interval over which the hop occurs. . . . [I]f we simply divide the action by Planck's constant [of action] we'll end up with a pure number. According to Feynman, this pure number is the amount we should wind the clock associated with a particle hopping from one place to another." The book specifies that one Planck constant or action unit of time corresponds to one complete twelve hour wind of an imaginary clock.

The imaginary clocks are postulated to be at all possible locations in the universe. Remember, the clocks are intended to simulate a wave that can fill all space. The little arrows [complex number vectors] representing the clock hand positions are combined in the same way described in the Feynman book for the stop watch hand positions. That is, the reader is assured that the mathematics will sum all the associated hand-pointing vectors over all the infinite possible paths the quantum could possibly travel in time and space to yield the *final arrow*, that is the predicted overall probability amplitude of the event.

The quantum universe book introduces a further rule: "we must shrink the clocks in addition to winding them counter-clockwise. The 'shrink rule' states that after all the new clocks have been spawned, every clock should be shrunk by the square root of the total number of clocks. . . . Of course there may be an infinite number of possible locations [there is as well as an infinite number of paths to reach those locations], which may sound alarming, but the maths can handle it." The book notes that the shrink rule is a simplification for discussion purposes in that some clocks get shrunk more than others because of Einstein's Special Theory of Relativity.

If the postulated displacement "is too close then the clock hands won't necessarily have the chance to go around at least once, which means that the clock hands won't necessarily have the chance to cancel each other so effectively [in the *orgy of quantum interference*]."

The book summarizes all this on "the premise that we must allow a particle the freedom to jump from any particular place in the Universe to absolutely anywhere else in an infinitesimally small moment."

The book continues to use its imaginary clock rules to show how the equation for the uncertainty principle follows—that is the rules for quantum mechanics probability waves—would alone yield the uncertainty principle—Heisenberg's use of unavoidable basic measurement uncertainty is not necessary to establish his uncertainty principle, only the wave motion of quanta. In doing so, it says: "[T]he precise amount by which we should turn the clock hand to account for the possibility that a particle hops a distance x in a time t is $mx^2/2ht$. .... The clock winding rule is obtained by associating each possible path the particle can take between two points, and it is an accident that after summing over all these paths, the result leads to this simple result [not including corrections to ensure consistency with Einstein's Theory of Special Relativity]."

Then it describes using elementary mathematics and its imaginary clock rules to derive the de Broglie equation, ρ [momentum] = h/λ that expresses the all-important "link between the property usually associated with particles—momentum—and a property usually associated with waves—wavelength." The book points out that, according to Fourier analysis, in order to get a wave packet corresponding to a particle it must consist of many wavelengths and thus momenta.

A further book section: "Interconnected" reasons, via the Pauli Exclusion principle, that every particle in the universe *knows* about the energy states of every other particle in the universe! It also discusses and illustrates Feynman diagrams, "the calculational tool of the professional particle physicist."

The book says: "It [QED theory] is single handedly capable of explaining all natural phenomena with the exception of gravity and nuclear phenomena. . . . certainly everything you see and feel around you." Then it asserts "that the weak and strong nuclear forces are described by exactly the same quantum field theoretic approach that we have described for QED. . . . Taken as a whole, the theory of these three forces is known, rather unassumably, as the Standard Model of particle physics."

Chapter Eight

The Quantum Universe moves onto discussing how electrons *orbiting* atomic nuclei are represented as standing waves in quantum theory. It analogizes standing waves to water waves confined in a swimming pool.

It later concludes that "quantum physics implies there is no such thing as empty space," it briefly discusses the Higgs particle that can explain why some fundamental properties have mass and remarks that: "Putting the theory to use led to the most important technological breakthrough of the twentieth century—the transistor."

In an epilogue: "The Death of Stars" the book discusses how the maximum mass of a dying [white dwarf] star has been determined by using the ratio $(h \times c/G)^{3/2} \times 1/m_p^2$, where in addition to the constants further discussed in this sequel, G represents Newton's gravitational constant and m_p represents the mass of a proton. The result of 1.4 times the mass of our sun is just under the maximum observed size of "around [the currently catalogued] 10,000 white dwarf stars."

References

P. C. W. Davies, *The Forces of Nature*, University of Cambridge, Cambridge University Press, 1979

Richard P. Feynman, *QED, The Strange Theory of Light and Matter*, Princeton, New Jersey, Princeton University Press, 1985

Brian Cox & Jeff Forshaw, *The Quantum Universe*, Boston Massachusetts, Da Capo Press, 2011

For further reading general augmenting QED material on the Web, you may click **Link 6** [http://en.wikipedia.org/wiki/Introduction_to_quantum_mechanics#Copenhagen_interpretation]
or **Link 7** [http://en.wikipedia.org/wiki/Quantum_electrodynamics] from the Web version of this book.

If you are interested in a mathematical treatment of the path integral procedure visit **Link 8** [http://en.wikipedia.org/wiki/Path_integral_formulation#Quantum_action_principle]

[When using the printed version of this book, to access the above links visit giac2002.org/qed_links.html and click the word "here" in the associated link number "Click here" phrase.]

Although not available at this book's initial writing, the following text is of potential interest:
Quantum Mechanics : A Modern Development (2nd Edition) Paperback – October 29, 2014 by Leslie E Ballentine (Author) ISBN-13: 978-9814578585